Moving the Earth

Recent Titles in
Contributions in Political Science

Watergate and Afterward: The Legacy of Richard M. Nixon
Leon Friedman and William F. Levantrosser, editors

Health Insurance and Public Policy: Risk, Allocation, and Equity
Miriam K. Mills and Robert H. Blank, editors

Public Authorities and Public Policy: The Business of Government
Jerry Mitchell, editor

Shepherd of Democracy: America and Germany in the Twentieth Century
Carl C. Hodge and Cathal J. Nolan, editors

Gerald R. Ford and the Politics of Post-Watergate America
Bernard J. Firestone and Alexej Ugrinsky, editors

The Democratic System in the Eastern Caribbean
Donald C. Peters

The Third World and South Africa: Post-Apartheid Challenges
Richard J. Payne

The Brazilian Legislature and Political System
Abdo I. Baaklini

Self-Determination in Western Democracies: Aboriginal Politics in a Comparative Perspective
Guntram F. A. Werther

United States Electoral Systems: Their Impact on Women and Minorities
Wilma Rule and Joseph F. Zimmerman, editors

Comparative Judicial Review and Public Policy
Donald W. Jackson and C. Neal Tate, editors

American Ambassadors in a Troubled World: Interviews with Senior Diplomats
Dayton Mak and Charles Stuart Kennedy

MOVING THE EARTH

Cooperative Federalism and Implementation of the Surface Mining Act

Edited by
UDAY DESAI

Prepared under the auspices of the Policy Studies Organization

Contributions in Political Science, Number 308

GREENWOOD PRESS
Westport, Connecticut • London

Library of Congress Cataloging-in-Publication Data

Moving the earth : cooperative federalism and implementation of the surface mining act / edited by Uday Desai ; prepared under the auspices of the Policy Studies Organization.
p. cm.—(Contributions in political science, ISSN 0147–1066 ; no. 308)
Includes bibliographical references.
ISBN 0–313–28698–1 (alk. paper)
1. Strip mining—Law and legislation—United States. 2. Strip mining—Law and legislation—United States—States. 3. Strip mining—Government policy—United States. 4. Federal government—United States. I. Desai, Uday. II. Policy Studies Organization. III. Series.
KF1823.M68 1993
346.7304'68—dc20
[347.306468] 92–25744

British Library Cataloguing in Publication Data is available.

Library of Congress Catalog Card Number: 92–25744
ISBN: 0–313–28698–1
ISSN: 0147–1066

First published in 1993

Greenwood Press, 88 Post Road West, Westport, CT 06881
An imprint of Greenwood Publishing Group, Inc.

Printed in the United States of America

The paper used in this book complies with the Permanent Paper Standard issued by the National Information Standards Organization (Z39.48–1984).

10 9 8 7 6 5 4 3 2 1

For my father, Chhabildas Desai,
who taught me the value of education

For my mother, Shantaben Desai,
who taught me the value of all things living

Contents

Acknowledgments

Editing a book is, I now fully realize, a difficult and time-consuming task. It is, in many ways, more difficult than writing one entirely alone. The editor incurs many debts, beginning perhaps with the authors of the various chapters in the book. I am fortunate to have convinced the nation's most knowledgeable and expert practitioners and academics on the implementation of the Surface Mining Control and Reclamation Act (SMCRA) to contribute to this book. They have been most cooperative and patient throughout the year and a half process of reviews, revisions, more reviews, and further revisions of their manuscripts. They have been prompt and good humored in their responses throughout. I am grateful to them all.

Special thanks are due to Richard Miller, who helped identify potential authors and reviewers, and read each chapter. He has been unflinching in his support and generous with his time and energy throughout preparation of this volume.

Editors and authors are fortunate when they can find a colleague who is willing to read and objectively comment on a manuscript. Reviewers receive no compensation for undertaking such a thankless task, save the knowledge that they have helped an (unknown) colleague and done a professional good deed to maintain a high quality of academic discourse. Many people have been generous with their time in reviewing manuscripts for this book. They include Ann Beier, Ann O'M Bowman, Greg Cawley, Cathy Carlson, Carl Close, John E. Freemuth, Michael S. Hamilton, D. Michael Harvey, Nina Rose Hatfield, Walter Heine, James Lester, Milo Mason, Patrick McGinley, Russell Price, Neal Shover, Al Whitehouse, and Karen Wiley. I am most grateful to them all.

I have also been fortunate to have had Mara Lou Hawse as copy editor for this book. She has been an exemplary copy editor. Every chapter in the book has benefited from her meticulous work. I am much indebted to her. The expert word processing of Olise Mandat and her staff has magically transformed so many "floppies" and their accompanying programs into one book manuscript. And she has managed this, I will never know how, most cheerfully. I am thankful to her and her staff.

John Mead and David Arey, director and associate director, respectively, of the Coal Research Center at Southern Illinois University at Carbondale provided the services of Ms. Hawse and Ms. Mandat. Without their help and generous support, this book would still be months away from completion. I am grateful to them both.

My wife, Christina, and my children, Michael, Kavita, and Neal, provided much needed support and care throughout my work on this volume. I will never know how to thank them adequately.

Moving the Earth

1 Policy Implementation and the Surface Mining Act: An Introduction

Uday Desai

The purpose of this book is to provide a perspective on policy implementation through a detailed analysis of experiences from one important environmental policy. This is a book about implementation of the Surface Mining Control and Reclamation Act of 1977. Its approach differs from that of most other texts in that not only do the contributors have widely differing points of view on the implementation of the Surface Mining Control and Reclamation Act (SMCRA), but also many of them are active participants in one or more aspects of its implementation. The authors are often involved in and committed to seeing SMCRA implemented in a fashion that is consistent with their own perspectives. This means that the views offered in this book will reflect the ambiguities and disagreements over both means and ends that attend any discussion of a significant public policy issue.

Some contributors offer the perspective of academic observers, presenting a scholarly discussion of the issues that surround the successes and failures of SMCRA. Others write from the perspective of involvement in the implementation struggle. It is our hope that this mix of views will help the student develop an appreciation of the complexities of policy implementation and come to understand that policy implementation does have consequences; it is a process of interaction among cohorts of engaged actors who have a commitment to making public policy work, albeit often to advance their own interests.

Even though it contains advocacy chapters, this is not an advocacy book. These chapters advocate opposing views. This is a book about concrete manifestations of theoretical formulations. It is an empirical case study of policy implementation.

PUBLIC POLICY PROCESS

Public policies are characterized by three principal attributes: that of purposiveness (they are constructed to achieve some end or set of ends); that of continuousness (they comprise a series of decisions and actions over a period of time); and that of formality (they are usually the consequence of governmental decision making and have as their basis some sort of formally enunciated decision which is, according to Mazmanian and Sabatier [1983, 20], "usually incorporated in a statute"). Public policies, however, are found in many places other than statutes. Executive orders, court decisions, and agency regulations all reflect public policy and direct courses of action toward achieving its ends. In modern government, regulations serve as the device for operationalizing the more general policy goals found in the statutes. In the case of SMCRA, the regulations that SMCRA has engendered are a major focus of policy development and implementation.

The most widely used construct of the public policy process is that of policy stages (Jones 1984; Anderson 1975; Peters 1986). Frequently the policy process is divided into five stages: agenda setting, formulation, legitimation, implementation, and evaluation. The policy stages model is essentially heuristic, not causal or prescriptive (Jones 1984, 28; Sabatier 1991, 147). But while the concept of stages is useful in organizing discussion and in understanding the dynamics of specific public policies and their consequences, it is important to note that in reality, the stages meld to form a seamless whole with many facets.

These different facets or stages are highly interrelated. Decisions, actions, and interactions at one stage or in one facet of the policy process are influenced by or tightly interrelated with decisions, actions, and interactions at another stage or in another arena of the policy process. Every public policy, in varying degrees, is continually and simultaneously formulated, legitimized, implemented, and evaluated. Each of these stages influences and overlaps other stages.

For example, policy formulation, the process by which national policy is made, has a profound effect on what will be achieved during implementation. A full explication of this process requires thorough understanding of the dynamics of governmental institutions (e.g., courts and bureaucratic agencies), the behavior of interest groups, networks and actors, and a knowledge of substantive policy. The successes and failures of policy solutions to any policy problem cannot be fully appreciated unless there is a clear understanding of all the forces that work to shape policy development.

Policy response to a problem is influenced by the contingencies of legitimation and by the way the problem has been defined. Problem definition shapes the parameters of possible policy responses, policy alternatives that are admissible in policy debate. It legitimates the policy participants and interests that will dominate the policy process. Agenda setting, the process by which policy problems and policy alternatives are defined and placed on the government agenda is not only complex, but also highly specific to each policy. Chance plays a significant role in agenda setting (Kingdon 1984). Contingencies of legitimation (i.e., of enacting a law) have direct and immediate impact on policy response to a problem. The need to achieve a majority coalition in Congress and to secure presidential assent in our fragmented and veto-ridden governmental process requires policy proponents to compromise with opponents and often leads to ambiguous or even contradictory policy response. Congress increasingly passes and the president signs into law public policies that instead of resolving the conflicts among various interests, move them from the legislative arena to the implementation arena. The ambiguities and contradictions in the statutes are deliberate, not inadvertent oversights by the lawmakers. They reflect the inability of the legislative process to resolve conflicts or to provide clear and specific guidance for carrying out legislative intents and purposes. The task of resolving conflicts among competing interests, of deciphering legislative intents and purposes, and of operationalizing legislative mandates, has increasingly been left to the implementation stage.

POLICY IMPLEMENTATION

The study of policy implementation has received much attention in the last 15 years or so. In the first edition of their now famous book *Implementation,* Pressman and Wildavsky (1973) were unable "to locate any thoroughgoing analysis of implementation" (xiii). By 1979, however, the second edition of the book called implementation studies "a growth industry" (1979,177). Dozens of books and hundreds of articles on policy implementation have been written and published. Implementation studies are considered to have entered a "third generation" (Goggin et al. 1990).

The reasons for this sustained interest in the study of policy implementation are varied and include the desire to understand an increasingly important and complex arena of public action and to learn about the hindrances to effective governmental actions in the field. Implementation studies generally "examine impediments to effective policy implementation and often identify why some policy legislation fails to be carried out in the way lawmakers had intended" (Desai 1989,3).

Many models, frameworks, conceptualizations, and theoretical perspectives of policy implementation have been advanced and discussed in the literature (for example, Smith 1973; Pressman and Wildavsky 1973, 1979; Van Meter and Van Horn 1975; Bardach 1977; Rein and Rabinovitz 1978; Elmore 1979–80; Nakamura and Smallwood 1980; Edwards 1980; Berman 1980; Mazmanian and Sabatier 1983; Linder and Peters 1987; Goggin et al. 1990; Hamilton 1990). It is difficult to neatly summarize the various approaches. However, they may be divided generally into two major groups: top-down and bottom-up (Desai 1991).

The top-down approach adopts the vantage point of the top policy maker(s) or central administrative authority. Its guiding normative assumption is the primacy of accomplishing the purposes of the central policy makers. Its principal concerns are control and compliance. It seeks to identify barriers to implementation and find regulatory and managerial strategies to overcome these barriers. A variation of this approach emphasizes the limitations of the control and compliance strategy and highlights the importance of environmental factors, such as intransigent groups and intractable problems. It stresses the need for clear and consistent policy objectives, precise legislative policy mandates, hierarchical integration, adequate fiscal resources, and sound causal policy theory.

The bottom-up approach is the view of policy implementation "from the trenches," from the field. It emphasizes the role of "street-level" actors in determining what actually happens "on the ground." It points out the severe limits on the ability of policy makers to accomplish their purposes. It focuses on the gap between the expressed policy intentions and actual accomplishments. Its fundamental point is that the actual policy (what really happens and is accomplished) is determined not by statutes containing clear and concise policy objectives and legislative mandates, nor by hierarchical control or adequate fiscal resources, but rather by bargaining between the street-level bureaucrats and their superiors and their clients or target groups. The premise of the bottom-up approach is that the policies and programs should be designed to be compatible with the preferences and predilections of the agency bureaucrats and their clients and target groups. Since they will control the outcome in any event, it would be wise to make policies explicitly to suit the agency's lower echelons and their clients.

A variation of this view makes interactions between various public and private actors in implementation its exclusive focus. It emphasizes that policy is not something decided and handed down by some central or higher authority, but rather it evolves in the process of implementation. The purposes and preferences of the central policy makers are largely irrelevant to implementation action.

Both the top-down and bottom-up views of policy implementation are generally pessimistic and see implementation as a quagmire for domestic policy objectives. The top-down proponents argue for tempering policy objectives to ensure "hardy" policies capable of meeting the harsh conditions for survival in implementation (unambiguous and consistent goals, clear mandates, hierarchical integration of implementors, sufficient fiscal resources, sound causal theory, etc.). Policies with ambitious but ambiguous goals or unproven causal policy theories are to be avoided. Fear of failure must discipline policy ambitions.

The bottom-up approach is even more pessimistic and critical of any significant domestic policy ambitions. It emphasizes the centrality of the interactions between public and private implementors, at the expense of the policy makers' intentions. Indeed, in this view, the more ambitious the intentions, the greater the disappointment. This approach argues for lowering our collective sights, for doing what we can do easily instead of trying to do what needs to be done. It counsels retreat on policy objectives. Its central criterion for selection of policy is ease of implementation. It puts policy implementation ahead of societal needs. It despairs of bringing about any significant change in social behavior through public policies and counsels against even trying. Over the last 20 years, studies taking the top-down or bottom-up view of policy implementation have identified many important variables (and categories of variables) in order to explain implementation behavior, that is, how (and, for some, how well) a public policy is implemented and, to a limited extent, why. Some analysts have emphasized political variables such as the power of organized interests (Murphy 1971; Van Meter and Van Horn 1975; Vietor 1980; Lindblom 1980; Ingram and Mann 1989); some have stressed the role of ideology (Fuchs 1988; Desai 1989) and party politics and competition (Van Meter and Van Horn 1975; Calvert 1989). Others have emphasized variables related to control (or lack thereof) of target groups and subordinate agencies and officials by the implementing agency and to compliance with agency directives. These variables have included financial resources and inducements and sanctions available to the implementing agency, hierarchical integration within and between the implementing agency and subordinate officials and agencies, and inter- and intra-organizational communications (Pressman and Wildavsky 1973; Edwards 1980; Nakamura and Smallwood 1980; Mazmanian and Sabatier 1983; Myrtle 1984; House and Shull 1985; Harris and Milkis 1989).

Still others have emphasized federalism and intergovernmental relations as the critical variables in understanding policy implementation (Derthick 1972; Ingram and Mann 1980; Conlan 1988; Vestal 1988; Scholtz and Wei 1986; Davis and Lester 1989; Wilds and Ganzel 1990; Desai 1992). A model that emphasizes federal-state-local relations has even been characterized as a "third generation" implementation model (Goggin et al. 1990).

Many additional variables have been identified over the years. The central role of the courts in influencing environmental policy implementation is now widely recognized (Melnick 1983; Wenner 1989; Galloway and FitzGerald 1985). Other variables include clarity and consistency of policy objectives and valid causal theory in the authorizing statutes, the implementing agency's commitment to policy objectives, the agency's leadership skills, technological difficulties and availability, characteristics of the target groups, overall socioeconomic conditions, and general public support for the policy objectives (Mazmanian and Sabatier 1983; Hamilton 1990).

THE SURFACE MINING CONTROL AND RECLAMATION ACT

This book is a case study. It is a study of the implementation of the Surface Mining Control and Reclamation Act of 1977, an important and highly conflictual environmental policy.

Unregulated surface mining is destructive to the environment. It causes erosion and, in steep mountainous terrain, even landslides, which in turn contribute to siltation, flooding, and water pollution with acidic and toxic runoff. It destroys fish and wildlife habitat and diminishes natural beauty. There are significant regional variations in the adverse environmental consequences of surface mining, depending upon the terrain, the hydrology, and the geological, biological, and chemical conditions of the mined area and upon the climate of the region.

In the mountainous Appalachian states, heavy rainfall erodes soil banks, producing mudslides and sedimentation, threatening homes, and choking off streams with silt and acidic runoff that cause flooding and kill aquatic life. In the generally flat midwestern states, the major problem is destruction of prime farmland and acidic runoff. The damage to farmland results from the mixing of higher quality topsoil horizons with minerals such as salt and pyrite found in lower soil horizons. Disturbed soil loses its original productivity. The damage to natural aquifers through contamination is the major problem in the western states. This region, with its low annual rainfall, is heavily dependent on its groundwater. The semiarid ecosystem of the region makes environmental recovery, whether by man through reclamation or by nature through natural succession, highly problematic.

By the early 1970s, more than four million acres of land in the United States had been affected by surface mining. In congressional hearings during the early 1970s, the evidence (including photographs) of severe damage to the environment—such as large areas of land churned into ruin by bulldozers, power shovels, and draglines, or streams flowing across state lines carrying mud from spoil heaps that settled in river beds and caused flooding—was presented by the advocates of a national policy to control surface mining. The expanding development of surface coal mining in the vast semiarid west and the fear of concomitant damage provided some urgency to act. Congress was presented with a problem of national scope that needed remedial action.

Environmental and coal state citizens groups and other advocates of strong federal regulation of surface coal mining argued that existing state regulations were weak and generally ineffective. They argued that an environmental problem of such national scope with a great impact on interstate commerce required a uniform national policy response to assure proper reclamation of surface mined land nationally and to equalize competition among coal industries in the different states. They wanted a strong federal regulatory role since they were highly suspicious of the ability or willingness of states to regulate surface mining (Shover et al. 1983).

Coal companies and industry trade groups opposed any federal regulation of surface coal mining. They argued that coal-producing states were already regulating the coal industry adequately and with more time would do an even better job. They also argued that terrain, hydrology, climate, and other conditions throughout the nation were diverse; therefore, a uniform federal law and regulations applicable to all states would be inappropriate and nearly impossible to implement.

Environmental and citizen groups fought long and hard (during five years and two presidential vetoes) for strong federal regulation of the surface coal mining industry with strict nationwide standards. The coal industry opposed any federal regulation or nationwide standards. The Surface Mining Control and Reclamation Act was signed into law by President Jimmy Carter in August 1977, after years of bitter fighting and acrimonious debate, in and out of Congress.

SMCRA established nationwide standards and a regulatory framework for the control of surface coal mining. The act, however, created a joint venture between the federal government and the coal states. It calls for "cooperation between" the secretary of the interior and the states to regulate surface coal mining. The act gives the "primary governmental responsibility for the development, issuance, and enforcement of surface mining regulations" to the states (Public Law 95–87, 101f). However, it also mandates federal standards and requires states to adhere to those standards. Framers of the act expected the coal states to develop their own regulatory programs in conformity with the federal act and regulations.

SMCRA created the Office of Surface Mining (OSM) in the Department of the Interior to carry out the responsibilities of administering the act and to help the states develop their own regulatory programs to "meet the requirements of the Act." SMCRA requires the secretary of the interior, acting through OSM, to review state programs, to determine whether they conform to the federal act and regulations. Once state regulatory programs are approved by the secretary, the states are granted "primacy"; that is, they then have the primary regulatory authority to control surface coal mining in the state. They are expected to administer and enforce the approved regulatory program. OSM's primary role then is oversight to ensure that the states indeed implement their approved programs.

OSM developed and promulgated the standards and regulations that the states were to follow. States were allowed to adapt, but not dilute, the national standards to their particular conditions. The act, recognizing the wide variations in regional conditions, provided a "state window" that allows states to experiment and develop regulations better suited to a specific climate, terrain, and other conditions. While they conform to national standards and regulations, the states are now the primary regulatory authority in all major coal-producing states. They approve mining plans, issue mining permits, inspect the mines, punish violators, and generally enforce the regulations. OSM, through its own inspections and other oversight activities, ensures that the states fully and conscientiously administer the act and enforce the regulations. The implementation of SMCRA essentially rested on the development and approval of the state regulatory programs by OSM and the granting of primacy to the states to administer and enforce their approved regulatory programs. Collectively, state regulatory programs would add up to a national program for surface coal mining control and reclamation.

ORGANIZATION OF THE BOOK

This book is about the implementation experience of an important environmental policy, as seen by its long-time students and by those directly engaged in and affected by its implementation. The majority of the chapters are written by those who have experienced and affected, firsthand, the implementation of SMCRA. They have been, and continue to be, involved in implementing the policy. The academic contributors have all been long-term scholars of SMCRA and its implementation. They have followed SMCRA's ups and downs with perhaps as much intensity as any of the practitioners.

This book does not outline yet another implementation framework or model. Rather, it is an implementation case study, not a conceptual or theoretical treatise. The chapters that follow do not flesh out any comprehensive implementation model. They do not deal with all the possible—or even all the important—variables that explain or have influenced SMCRA implementation behavior. Instead, they focus on three critical variables that have shaped SMCRA's implementation.

The role of various interests, economic interests especially, has been given little attention in implementation frameworks and in studies on the implementation of specific policies, even though their role is widely recognized in policy agenda setting, formulation, and legitimation (Lindblom 1977). Three chapters in Part II deal with three major interests directly affected by SMCRA. Harold P. Quinn, Jr., in chapter 4, provides the perspective on implementation of SMCRA from the coal industry. He argues that forces more fundamental than SMCRA—technological and market forces—are at work and have significantly changed the structure of the coal industry in recent decades. To develop relevant regulatory policy, those responsible for implementing SMCRA must understand and take into account these fundamental forces and their impacts on the coal industry. James M. McElfish, Jr., in chapter 5, provides the perspective of environmental citizen groups. He describes the strategies used by the groups in their efforts to affect implementation of SMCRA and to preserve the gains made during the legislative process. He argues that during the 1980s environmental and citizen groups saw OSM as hostile to their agenda. In response, these groups actively resorted to the courts and Congress to influence the course of SMCRA's implementation. In chapter 6, Gregory E. Conrad, representing state interests, provides a forceful expression of the states' perspectives on SMCRA's implementation. He argues that OSM in its implementation of SMCRA has been less than faithful to the federal-state partnership embodied in the act. Conrad is hopeful that the balance generally in the roles of federal and state governments in policy implementation may well be in the process of being readjusted to provide a more active role for states. The coal industry, environmental groups, and the states have been the three major interest groups that have actively and consistently engaged in the conflicts over

SMCRA's implementation. These three chapters provide an account of the arguments and strategies of each group and their impacts, as seen by the groups themselves.

The role of the courts in policy implementation, especially the implementation of environmental policies, has been both massive and critical. Although implementation models and frameworks do not generally identify courts as an especially important variable, many implementation case studies of specific policies have recognized their critical role, and many long-time participants and observers of SMCRA have recognized the especially critical and continuing role of the courts in its implementation. They characterize SMCRA's implementation experience as "litigating the agency (OSM) and regulations in place." SMCRA has been litigated, perhaps more than most other environmental policies. Before SMCRA was enacted, its opponents declared that it would invite "endless litigation" ("Strip Mining Control Bill Signed" 1977, 623), and indeed it has. From broad constitutional attacks to continuing court challenges to its specific regulations, lawsuits have mounted against SMCRA since its passage in 1977. It is impossible to understand SMCRA's implementation without a careful and direct look at the role of the courts. Chapters 7 and 8 in Part III address this issue. In chapter 7, Robert E. Beck provides a detailed account of the role of the federal courts in implementing SMCRA. He observes that court decisions underscore the authority of the implementing agency to fashion SMCRA's regulatory scheme. This authority significantly constrains the capacity of various interest groups to affect policy implementation. Lisa J. Morris, in chapter 8, focuses on the consequences of the continuing court challenges and court-ordered modifications in regulations on the implementation of SMCRA on the ground at the state level. She argues that these litigations have created a continuing instability in state regulatory programs and have gravely undermined the effectiveness of SMCRA's implementation in the coal states.

Intergovernmental relations and the federal nature of our political system are increasingly seen as the critical and the most important variables in understanding and explaining policy implementation in the United States. Indeed, recently a focus on intergovernmental implementation has been advanced as the model for third-generation implementation research (Goggin et al. 1990). The critical importance of federalism and intergovernmental dynamics in explaining domestic policy implementation behavior in the United States has, of course, long been recognized (Murphy 1971; Derthick 1972). Increasingly, environmental health and safety policy statutes have specified state "primacy" or "partial preemption" as the implementation strategy. This tendency has once again focused the attention of students of policy implementation on intergovernmental relations and the role of the states in policy implementation.

SMCRA, as discussed earlier, requires intergovernmental implementation. It gives primacy to the states to implement the act. Thus, the states are at the center of implementing SMCRA. To understand and explain SMCRA implementation behavior is to understand its implementation in the states. The largest section of the book, Part IV, therefore is devoted to state-level implementation. John C. Dernbach, in chapter 9, details the experiences of Pennsylvania in implementing SMCRA. He argues that SMCRA has resulted in improved technical competence and better staffing of the state's regulatory program, and the net effect has been to improve the regulation and reclamation of surface coal mining in Pennsylvania. He also shows that SMCRA's detailed requirements have created serious difficulties in dealing with several issues, especially bonding, that were largely left unaddressed by SMCRA. In essence, he argues that the very specificity and detailed nature of the act have become hindrances in dealing with issues that were unanticipated or left unattended during the legislative process.

Denise A. Dragoo, in chapter 10, discusses the implementation of SMCRA in the western coal states. She finds that implementation of SMCRA by OSM has favored national uniformity based on regulations designed primarily for conditions in eastern coal states.

Intergovernmental implementation assumes that states are willing and able to implement the policy. Three chapters in Part IV empirically explore this assumption in the case of SMCRA. Michael J. Scicchitano, David Hedge, and Patricia Metz, in chapter 11, present quantitative data on the variations in implementation performance of eastern, midwestern, and western states. Denise Scheberle, in chapter 12, presents quantitative data on the differences in perceptions about state capabilities and performance in implementing SMCRA, among state and federal inspectors and field level personnel, that is, among those who are engaged in day-to-day implementation activities and whose cooperation and respect for each other's competence and commitment are critical to effective implementation of SMCRA. She finds a wide gulf between the perceptions of state and federal implementors. Kathryn M. Mutz and Ronald W. Daniels directly address the issue of state capabilities in chapter 13. They conclude, based on extensive data collected from all but one coal state, that the design of SMCRA—intergovernmental implementation design—has resulted in improved capabilities for most states that had deficient abilities before SMCRA, while it has not made any significant difference in the states that had high capabilities.

Two chapters in Part I, following this introduction, provide general background on SMCRA and OSM. In chapter 2, Richard O. Miller first provides a brief legislative history of the act, then he discusses the consequences of three specific examples of legislative ambiguity on the implementation of SMCRA by OSM. Miller concludes that the legislative ambiguities have retarded SMCRA's comprehensive implementation. Donald C. Menzel, in chapter 3, provides a brief history of the OSM since its inception in 1977. It is a checkered history. It is a profile of an agency in constant turmoil, still struggling to find a steady balance amid the competing interests and ideological preferences of those affected by it.

REFERENCES

Anderson, James. 1975. *Public Policy-Making*. New York: Praeger.

Bardach, Eugene. 1977. *The Implementation Game: What Happens After a Bill Becomes a Law*. Cambridge, Mass.: MIT Press.

Berman, Paul. 1980. "Thinking about Programmed and Adaptive Implementation: Matching Strategies to Situations." In Helen M. Ingram and Dean E. Mann (eds.), *Why Policies Succeed or Fail*. Beverly Hills, Calif.: Sage.

Calvert, Jerry W. 1989. "Party Politics and Environmental Policy." In James P. Lester (ed.), *Environmental Politics and Policy: Theories and Evidence*. Durham, N.C.: Duke University Press.

Conlan, Timothy. 1988. *New Federalism: Intergovernmental Reform from Nixon to Reagan*. Washington, D.C.: The Brookings Institution.

Davis, Charles E., and James P. Lester. 1989. "Federalism and Environmental Policy." In James P. Lester (ed.), *Environmental Politics and Policy: Theories and Evidence*. Durham, N.C.: Duke University Press.

Derthick, Martha. 1972. *New Towns In-Town*. Washington, D.C.: Urban Institute.

Desai, Uday. 1991. "Implementation of Public Policy: Note on American Experience." Paper presented at research conference, "Challenges of Public Administration," sponsored by Southern Illinois University at Carbondale and Land Vorarlberg, Scholb Hofen, Landes-Bildungszentrum, Austria, June.

———. 1992. "The Politics of Federal-State Relations: The Case of Surface Mining Regulations." *Natural Resources Journal* 31: 141–50.

———. 1989. "The Politics of Implementation Design." *Journal of Urban and Contemporary Law* 36: 3–19.

Edwards, George C., III. 1980. *Implementing Public Policy*. Washington, D.C.: C. Q. Press.

Elmore, Richard F. 1979–80. "Backward Mapping: Implementation Research and Policy Decisions." *Political Science Quarterly* 94: 601–16.

Fuchs, Edward Paul. 1988. *Presidents, Management and Regulation.* Englewood Cliffs, N.J.: Prentice-Hall.

Galloway, L. T., and T. J. FitzGerald. 1985. "Abuse of the Surface Mining Act: A Continuing Story." *West Virginia Law Review* 87: 627–47.

Goggin, Malcolm L., Ann O'M Bowman, James P. Lester, and Laurence J. O'Toole, Jr. 1990. *Implementation Theory and Practice Toward a Third Generation.* Glenview, Ill.: Scott, Foresman and Company.

Hamilton, Michael S. 1990. "Introduction." In Michael S. Hamilton (ed.), *Regulatory Federalism, Natural Resources, and Environmental Management.* Washington, D.C.: American Society for Public Administration.

Harris, Richard A., and Sidney M. Milkis. 1989. *The Politics of Regulatory Change.* New York: Oxford University Press.

House, Peter W., and Roger D. Shull. 1985. *Regulatory Reform: Politics and the Environment.* Lanham, Md.: University Press of America.

Ingram, Helen M., and Dean E. Mann. 1989. "Interest Groups and Environmental Policy." In James P. Lester (ed.), *Environmental Politics and Policy: Theories and Evidence.* Durham, N.C.: Duke University Press.

Ingram, Helen M., and Dean E. Mann (eds.). 1980. *Why Policies Succeed or Fail.* Beverly Hills, Calif.: Sage.

Jones, Charles D. 1984. *An Introduction to Public Policy.* 2d ed. North Scituate, Mass.: Wadsworth Publishers.

Kingdon, John W. 1984. *Agendas, Alternatives, and Public Policies.* Boston: Little, Brown.

Lindblom, Charles E. 1977. *Politics and Markets: The World's Political-Economic Systems.* New York: Basic Books.

———. 1980. *The Policy-Making Process.* 2d ed. Englewood Cliffs, N.J.: Prentice Hall.

Linder, Stephen M., and B. Guy Peters. 1987. "A Design Perspective on Policy Implementation: The Fallacies of Misplaced Prescription." *Policy Studies Review* 6: 459–75.

Mazmanian, Daniel A., and Paul A. Sabatier. 1983. *Implementation and Public Policy.* Glenview, Ill.: Scott, Foresman and Company.

Melnick, R. Shep. 1983. *Regulation and the Courts: The Case of the Clean Air Act.* Washington, D.C.: Brookings Institution.

Murphy, Jerome. 1971. "Title I of ERISA: The Politics of Implementing Federal Education Reform." *Harvard Educational Review* 41: 35–49.

Myrtle, R. C. 1984. "A Managerial View of Policy Implementation." *American Review of Public Administration* 17: 17–32.

Nakamura, Robert T., and Frank Smallwood. 1980. *The Politics of Policy Implementation.* New York: St. Martin's Press.

Peters, B. Guy. 1986. *American Public Policy: Promise and Performance.* 2d ed. Chatham, N.J.: Chatham House.

Pressman, Jeffrey, and Aaron Wildavsky. 1973. *Implementation.* Berkeley: University of California Press.

———. 1979. *Implementation.* 2d ed. Berkeley: University of California Press.

Rein, Martin, and Francine Rabinovitz. 1978. "Implementation: A Theoretical Perspective." In W. Burnham and W. Weinberg (eds.), *American Politics and Public Policy.* Cambridge, Mass.: MIT Press.

Sabatier, Paul A. 1991. "Toward Better Theories of the Policy Process." *PS Political Science and Politics* 24: 147–156.

Scholtz, John T., and Heng Wei. 1986. "Regulatory Enforcement in a Federal System." *American Political Science Review* 80: 1249–70.

Shover, Neal, John Lynxwiler, and Donald Clelland. 1983. *Developing a Regulatory Bureaucracy: The Office of Surface Mining Reclamation and Enforcement.* Washington, D.C.: National Institute of Justice.

Smith, T. B. 1973. "The Policy Implementation Process." *Policy Sciences* 4: 197–209.

"Strip Mining Control Bill Signed." 1977. *Congressional Quarterly Almanac* 33: 617–23.

Van Meter, Donald S., and Carl E. Van Horn. 1975. "The Policy Implementation Process: A Conceptual Framework." *Administration and Society* 6: 445–88.

Vestal, Theodore. 1988. "Federal Administration of the Surface Mining Control and Reclamation Act of 1977 in Oklahoma." *Publius* 18: 45–60.

Vietor, Richard. 1980. *Environmental Politics and the Coal Coalition.* College Station: Texas A & M University Press.

Wenner, Lettie McSpadden. 1989. "The Courts and Environmental Policy." In James P. Lester (ed.), *Environmental Politics and Policy: Theories and Evidence.* Durham, N.C.: Duke University Press.

Wilds, Leah, and Richard Ganzel. 1990. "The Changing Face of Regulatory Federalism: Decentralization—or Nationalization." In Michael S. Hamilton (ed.), *Regulatory Federalism, Natural Resources, and Environmental Management.* Washington, D.C.: American Society for Public Administration.

I Legislative and Administrative Legacy

2 The Surface Mining Control and Reclamation Act: Policy Structure, Policy Choices, and the Legacy of Legislation

Richard O. Miller

INTRODUCTION

The purpose of this chapter is to outline briefly the legislative history of the Surface Mining Control and Reclamation Act (SMCRA) and identify important policy and structural choices made and not made by Congress. Clear legislative policy choices have obvious consequences for implementation, and when the product of the legislative process is ambiguity or non-choices, there are equally obvious consequences for implementation. Non-choices—gaps in the fabric of legislative intent—create or contribute to a policy vacuum that can plague implementing agencies. Coping with these omissions is often the most troublesome aspect of implementation. This is doubly true where policy implementation is carried out within a framework of cooperative federalism: where state or local partners in the implementation process must either wait until a policy gap is filled by the federal agency or undertake the risky and costly task of devising and putting into place their own solutions. This chapter identifies three separate legislative legacies of SMCRA and traces how these legacies have affected implementation of the act.

THE LEGISLATIVE HISTORY IN BRIEF

"We have here today an old friend . . . the surface mining bill." Thus, Congressman Morris Udall introduced what had for him become a legislative priority. Indeed it was an old friend (*Congressional Record,* House, May 7, 1975, 13368). As early as 1940, Representative Everett Dirksen had introduced legislation that would have provided for federal regulation of surface mining (*Legislative Proposals* 1971).

Following Dirksen's initial effort in 1940, federal surface mining legislation continued to be introduced periodically over the next 30 years. However, it was 1971 before Congress exhibited sustained interest in enacting legislation to control the effects of surface mining. It is testimony to the political intractability of many of the issues surrounding regulation of surface mining that, despite repeated efforts, several more years of struggle followed before Congress enacted, and the president signed, legislation that established a comprehensive federal program to regulate surface mining.

SMCRA was the culmination of more than six years of focused effort to develop federal policy for surface mining. From 1971, when Representative Ken Hechler of West Virginia introduced legislation that would have prohibited surface mining entirely, to SMCRA's final passage in 1977, congressional attempts to devise national surface mining policy highlighted how Congress and interest groups interact to forge a sustainable consensus on controversial issues.

Developing policy for surface mining was not easy for Congress, nor was the final legislation succinct or dispositive. Indeed, the final version of SMCRA did not do everything that ought to be done, and in the view of the participants in SMCRA's development, it did many things that ought not to have been done. One Congressman observed that the final conference report, filed on July 12, 1977, contained many compromises, both major and minor (*Congressional Record,* House, 1977, 7589). Another, urging rejection of the conference report, speculated that "this regulatory measure seems destined to become a bureaucratic nightmare" (*Congressional Record,* House, 1977, 7589).

Rather than a vivid legislative win for proponents of federal regulation of surface mining, SMCRA appears to be instead a tally sheet in which various interest groups with a stake in surface mining policy won on some issues, lost on others, and deferred resolution of still others.

The 92nd Congress

Initial legislation that would have outlawed surface mining attracted the support of nearly one hundred cosponsors. Other bills, including an administration bill, varied as to the minerals and activities they proposed to regulate. The administration proposal (H.R. 5689), for example, would have required the states to regulate all mining and reclamation operations. Other proposals differed in their provision for federal funds for reclamation and restoration of previously mined areas, regulation of activities on federal lands, and the extent of citizen participation. H.R. 6482, the proposal introduced by Congressman Wayne Hays of Ohio, became the surface mining bill of the 92nd Congress. H.R. 6482 addressed only surface coal mining. The bill was passed by the House on October 11, 1972, by a vote of 265 to 75 (*Congressional Record,* House, 92nd Congress, 2nd Session, 35046).

Companion legislation was introduced in the Senate, along with several other related bills, and Senate hearings were held in 1971. However, no bill was enacted. At this point, significant differences existed between the House and Senate approaches to mining policy. The Senate legislation, S. 630, provided for federal regulation of all mining, rather than surface mining exclusively. Both House and Senate versions provided for primary regulation to be the responsibility of the federal government.

The 93rd Congress

In the House, draft strip mining legislation was referred to the House Interior Committee, where it was jointly referred to the Mines and Mining Subcommittee and to the new Subcommittee on the Environment chaired by Representative Morris Udall of Arizona. The bill, known as H.R. 11500, was reported out of the Interior Committee on May 14, 1974. The bill was debated for six days on the floor of the House, and more than one hundred amendments were considered. On July 25, 1974, the House passed H.R. 11500 by a vote of 291 to 81 (*Congressional Record,* House, 93rd Congress, 2nd Session, July 25, 1974, 6705).

In the Senate, Senator Henry Jackson, chairman of the Senate Interior Committee, introduced S. 425. A major difference in this Senate legislation was that it focused on regulation of coal mining in contrast to earlier versions of the Senate bill, which addressed all minerals. The Senate passed S. 425 on October 9, 1973. Following extended debate, the conference report was passed by both houses in December, 1974, more than a year later, and "The Surface Mining Control and Reclamation Act of 1974" was sent to President Gerald Ford. The president pocket-vetoed the measure on December 30, 1974.

The 94th Congress

In light of the support voiced for the vetoed legislation and the substantial majorities that had voted for the passage of the vetoed legislation, the surface mining act was reintroduced as H.R. 25 in the House and as S. 7

in the Senate in the 94th Congress. Some observers expected that the legislation would be easily passed in that Congress. The administration outlined objections to the bill, and in February various administration officials testified before the House Interior Committee as to the changes needed. The bill, reflecting to some extent the changes requested by the administration, was reported out of committee on February 27, 1975. The House debated the bill in mid-March, adopted 22 amendments, and passed the bill on March 18, 1975, by a vote of 333 to 86 (*Congressional Record*, House, 94th Congress, 1st Session, 1908).

The House and Senate versions contained significant differences and the Administration continued to press for further changes. The conference report was passed by both bodies and, given the 293 to 115 majority in the House, appeared veto-proof. However, President Ford vetoed the bill on May 20, 1975. Surprisingly, the House fell three votes short of overriding the veto, primarily because Republican members supported the president on the veto vote (*Congressional Record*, House, 94th Congress, 1st Session, 5205).

In his veto message, President Ford justified his action on the grounds that the legislation would restrict coal production, increase dependence on Middle East oil, raise consumer prices, and increase unemployment. The data upon which the veto was based were subject to varying interpretations and controversy. Chairman Udall went so far as to hold hearings on the veto on June 3, 1975.

The surface mining bill was reintroduced in the second session of the 94th Congress. However, the House Rules Committee voted to table further consideration of the bill, and no action was initiated during the remainder of the 94th Congress.

The 95th Congress: The Enactment of SMCRA

The surface mining bill was reintroduced in the 95th Congress. President Carter had indicated that he would approve coal strip mining legislation during his electoral campaign, and there was considerable momentum to produce legislation for the president to sign during his first year in office.

H.R. 2 was introduced in the House and the Subcommittee on Energy and the Environment of the House Interior and Insular Affairs Committee received testimony on January 10 and 12, 1977. Testimony dealt with various surface mining issues in preparation for later hearings. The House Subcommittee held further hearings on the bill in February. On April 22, 1977, the House Committee on Interior and Insular Affairs recommended passage of H.R. 2. The Subcommittee on Public Lands and Resources of the Senate Committee on Energy and Natural Resources held hearings on S. 7 in February 1977.

On May 10, 1977, the Senate Committee on Energy and Natural Resources favorably reported out S. 7. After passage in the Senate and a brief Conference, the House completed its consideration of the conference report on June 30, 1977. The Report was filed on July 12, 1977, and was agreed to by the House on July 21, by a vote of 325 to 68. Following concurrence by the Senate, the legislation was signed on August 3, 1977, by President Jimmy Carter, in a Rose Garden ceremony.

THE "POLICY LEGACIES" OF THE LEGISLATIVE PROCESS

The legislative struggle can be succinctly characterized as a series of small victories for proponents and opponents of federal surface coal mining regulation. The Surface Mining Act and the agency it created are excellent examples of the compromises and contradictions that attend any major policy formation effort in a pluralist society.

SMCRA is certainly no testament to optimal policy formation, if one's definition of optimal policy formation encompasses relatively "clean" and concise statutory language, nor is the implementing structure it created a model for bureaucratic efficiency and rationality, if economy, clear lines of authority, and "rational" organizational structure are important criteria.

In opposing a committee report on an immediate predecessor to SMCRA, one critic speculated that enactment of surface mining legislation would "lead to countless years of regulatory delays, litigation and uncertainty against the best interest of achieving our environmental objectives because of ambiguous, vague, and complex provisions" (U.S. Congress, House 1976, 142).

The act and OSM are the culmination of a series of compromises among opposing groups. Some groups wished the act and the agency to have maximum reach. Other groups sought to limit the implementation of a national surface mining policy and to constrain a federal agency in its implementation of the policy. "Opponents," argues Terry M. Moe, "want structures that work against effective performance. They fear strong, coherent, centralized organization. They like fragmented authority, decentralization, federalism, checks and balances, and other structural means of promoting weakness, confusion, and delay" (1989, 276).

In the case of SMCRA, industry groups opposed direct federal regulation of surface coal mining, and instead urged that states be given the lead for controlling surface coal mining. In testimony on earlier bills, a coal industry representative offered the industry's view that "the approach . . . which encourages the States to develop their own programs based on broad Federal criteria, provides the most effective way of insuring [the] objective [of sound reclamation]" (U.S. Congress, Senate 1971, 329). Industry efforts in this regard were successful, as the concept of a pronounced state role for regulating surface mining was found in most versions of strip mining legislation. Environmental groups were quick to characterize industry opposition to a centralized regulatory scheme as a device to weaken the regulatory effort.

States, for the most part, also opposed a predominant federal role. The congressional movement to pass surface coal mining legislation stimulated many coal mining states to pass or to amend their existing legislation regulating surface mining. Although the motive for such actions is somewhat speculative, it can be argued that state legislative activity during this time was in part due to a desire to head off federal regulation. Congressmen who opposed federal legislation introduced evidence that most states were actively initiating or updating legislation to improve regulation of surface mining. State officials also expressed the view that, in the main, their regulatory programs, often the subject of recent strengthening, were working satisfactorily (U.S. Congress, House 1976, 143).

Environmental groups were suspicious of the ability of the states to implement a rigorous program for controlling surface mining and were well aware that industry could expect to obtain a better hearing for its views in most coal states. These groups urged Congress to adopt a centralized approach to regulation.

SMCRA is more than a series of policy compromises. It also includes a series of structural compromises that may advance or retard policy implementation. To paraphrase Eugene Bardach, structural choices and their implementation are "the continuation of politics by other means" (Bardach 1977). Particularly with regard to its basic structure—where in the federal government an agency is located and with whom it must transact—OSM's location was viewed as an important issue by participants in the legislation that became SMCRA. As Moe observes,

> The choices about structure that are made [during initial legislation] when the agency is designed and empowered with a mandate, are normally far more enduring and consequential than those that will be made later. They constitute an institutional base that is protected by all the impediments to new legislation inherent in separation of powers. . . . Most of the pushing and hauling in subsequent years is likely to produce only incremental change. (1989, 285)

Indeed, OSM has remained as originally placed by SMCRA in the Department of the Interior. Occasional suggestions to move it to other institutional locales have not been acted upon. This location was not the choice of all involved in SMCRA's development. In urging adoption of one of SMCRA's forerunners, Congressman Miller of Ohio expressed the view that he "would have preferred to see EPA given jurisdiction" (*Congressional Record,* House, 92nd Congress, 2nd Session October 11, 1972, 35052). A leading advocate of banning surface mining outright expressed his conclusions that "any legislation which places administrative control over strip mining in the Department of Interior, rather than the Environmental Protection Agency, isn't worth the paper it's written on" (U.S. Congress, Senate 1977, 702). In assessing the progress of surface mining legislation, one observer concluded:

It is unclear whether the opponents or proponents of the legislation have gained more from the repeated veto actions and continuing consideration of the Federal Strip Mine bill. It appears, however, that each time Congress reassesses this legislation those additional variances granted to the industry are balanced by more precise specificity in both language and legislative history, thus neutralizing industry enthusiasm for the legislation, if not stimulating increased opposition to it. (Dunlap 1975, 29)

As final consideration of S. 7 moved forward, it became clear that SMCRA would contain more than its share of ambiguities. Such ambiguity may often be the price of legislative agreement, but agreement sufficient to secure passage of legislation often merely defers conflict or delays resolution of difficult policy choices. Senator Lee Metcalf, the floor manager of S. 7, in responding to the suggestion of ambiguity in one of the act's terms replied:

I have frequently been exposed to alleged ambiguity. I found out the courts are well equipped to interpret and to give a construction to some of the most loose language the Senate has ever written into legislation. . . . I doubt if there is any ambiguity that cannot be interpreted either by Secretarial regulations or by an appeal to an appropriate court. (*Congressional Record,* Senate, May 19, 1977, 8039)

Indeed, much of the activity surrounding the regulatory program provisions of the act has concerned the "fleshing out" of the language of the act with more detailed regulations. This has been a lengthy process for OSM and its state partners. The success with which the scheme has been "fleshed out" may be a function of the character of the ambiguities in the act and whether the gap is one of a conceptual nature or is more attributable to the political impossibility of developing a consensus. It may also be true that the source of ambiguity—whether it is a function of technical complexity or is an indication of the lack of a political or policy consensus, may contribute to the agency's success or lack of success in bringing to closure issues left open by the legislative process.

In the examples that follow, each issue represents an instance where Congress left an issue for the implementing agency to resolve. The first, the definition of Valid Existing Rights, presents a situation where development of a workable definition of Valid Existing Rights does not seem insurmountable. However, for Congress, and certainly for the implementing agency, the development of a political consensus necessary to sustain any given definition has been elusive. The second issue, the degree to which a definition of farming would preclude mining in alluvial valley floors, appears to be technically difficult, but the possibility of a political consensus being developed appears to have been more feasible. The final example, dealing with the larger issue of the federal and state roles in a cooperative federalism scheme, appears to present both technical and political elements that are difficult to surmount in a dispositive fashion.

VALID EXISTING RIGHTS: CONCEPTUALLY DIRECT DEFINITION, AND LACK OF POLITICAL CONSENSUS

Section 701 of SMCRA contains an extensive list of definitions, ranging from what constitutes a surface coal mining operation to what is meant by the term prime farmland. However, Congress failed to define what was to become a major point of struggle in implementing SMCRA. What, precisely, constitutes a valid existing right to mine coal?

The term Valid Existing Rights (VER) defines an exception to the prohibitions on surface coal mining in areas enumerated in section 522(e) of SMCRA. The areas include national parks, national forests, public buildings, cemeteries, and churches. Instead of an outright ban, Congress provided that, subject to "valid existing rights," surface coal mining operations were prohibited in the enumerated areas. However, the failure of Congress to provide a definition has contributed to 13 years of policy ambiguity and uncertainty. Although the reasons for the omission are not possible to identify with certainty, "It is most likely that Congress really didn't know what was meant by the term" (McElfish 1989–90, 624).

OSM first attempted to define VER in its 1979 permanent regulatory program regulations. In that rule making, OSM adopted the position that VER existed where an applicant had obtained all necessary permits to mine coal prior to the date of SMCRA's enactment. The VER definition was among the elements in the permanent program regulations that were challenged by both industry and environmental groups. The definition was rejected by the court in a decision of the Washington, D.C., District Court, In re Permanent Surface Mining Regulations litigation, heard by Judge Thomas Flannery. Judge Flannery held that such a definition was likely too narrowly drawn and suggested that making "a good faith effort" to obtain all necessary permits might suffice to qualify for VER.

Pending further rule making, the agency suspended the definition and essentially adopted as policy the court's suggestion that a good faith effort to obtain all needed permits would suffice to establish VER. For all intents and purposes, the good faith standard became the standard for state implementation of the act. State programs were adopted or amended to reflect the different interpretation, pending further OSM rule making on the issue. The process of amending state programs can require one to two years and may also require action by state legislatures. This process was moving forward as OSM initiated another rule making on VER.

In 1983, the OSM permanent program regulations were substantially revised as part of the Reagan administration's effort at regulatory reform. This included proposing a new definition of VER, this time to include the "good faith effort" to obtain all permits standard suggested by the court in the earlier litigation. However, in the final rule making, OSM adopted a revised VER definition that included the concept of "takings." The definition stipulated that where denial of VER would constitute a compensable taking under the fifth and fourteenth amendments, VER would be determined to exist.

The 1983 revisions to the permanent program regulations were, in turn, litigated. In 1985, the court held that the VER definition adopted in 1983 departed substantially from the regulations that were proposed, and failure to adequately advise the public that OSM was considering such a definition violated the notice and comment provisions of the federal Administrative Procedure Act. In 1986, OSM suspended the 1983 regulation, to the extent that it included a "takings" component. This left state regulatory authorities with guidance for determining VER which was, at a minimum, less than secure. Critics argue that the 1986 suspension resulted in a definition being used that *also* had never been properly promulgated in compliance with the Administrative Procedure Act.

Two years after publishing the 1986 suspension notice, OSM proposed a new regulatory definition for VER. The proposed options included the "good faith all permits" test and a more permissive "ownership of the coal resource plus the right to mine by the method intended" standard. The public comment period lasted 118 days, and OSM received approximately forty-four hundred comment letters from the public. In response to the comments and a barrage of generally negative publicity, the proposed rule was withdrawn on July 18, 1989. It would require an additional two years for OSM to propose a new definition of VER, one that incorporated the "good faith effort" and the "takings" standards.

The most fundamental consequence of the creation of a policy gap by the congressional failure to define VER has been an unstable regulatory program. From a legislative perspective however, there may exist some very sound reasons for opting for ambiguity. The VER rule has been controversial each time it has been proposed. Thirteen years of regulatory and legal struggle have yet to provide a dispositive definition of the protection required for the areas enumerated in section 522(e) of SMCRA.

From the perspective of implementation, the history of VER under SMCRA also stands for the proposition that agencies will have no easier task than Congress in forging a policy consensus on an issue where a consensus eluded the legislature. Further time and resources expended in attempting to do so will be considerable.

PROHIBITIONS ON MINING IN ALLUVIAL VALLEY FLOORS: A TECHNICAL SOLUTION TO A CONCEPTUALLY DIFFICULT LEGACY

There are many instances in SMCRA where legislative language has been adopted to specifically address issues that have regional or local consequences. A case in point is the congressional attempt to address how and to what extent surface mining should be prohibited or regulated in Alluvial Valley Floors (AVFs) west of the 100th meridian.

Congress was well aware that significant coal resources underlay the vast intermountain valleys of the West, particularly in Colorado, Utah, Wyoming, and Montana. At the same time, these valleys play a key role in the ranching and farming economies of the arid West. Congress attempted to protect those areas that are "significant" to farming and ranching.

Section 701 of the act defines Alluvial Valley Floors (AVFs) as

> the unconsolidated stream laid deposits holding streams where water availability is sufficient for subirrigation or flood irrigation agricultural activities but which does not include upland areas which are generally overlain by a thin veneer of colluvial deposits composed chiefly of debris from sheet erosion, deposits by

unconcentrated runoff or slope wash together with talus, other mass movement accumulation or windblown deposits.

In turn, section 510(b)(5) of the act requires that no permit or application for a surface coal mine shall be approved unless the information demonstrates that the proposed surface coal mining operation, if located west of the 100th meridian, would "not interrupt, discontinue, or preclude farming on alluvial valley floors that are irrigated or naturally subirrigated . . . or not materially damage the quantity or quality of water in surface or underground water systems that supply these valley floors." Thus, SMCRA provides varying degrees of protection to AVFs depending on whether they are significant to farming activities. If they are, surface mining is prohibited where such operations would "interrupt, discontinue, or preclude" farming.

OSM's 1979 permanent program regulations did not define farming. In 1983, as a result of regulatory reform, the definition of agricultural activities was revised to include farming. This meant that more AVF lands would have been placed off limits to mining. Industry challenged this definition as being overly broad and inconsistent with the act and with congressional intent. The industry challenge was sustained by the courts, and OSM subsequently suspended the definition in 1985. The court also stated that it "would not presume to define these terms" (agricultural activities and farming), leaving it to the agency to develop a definition that reflected legislative intent.

In 1988, OSM proposed revised definitions and began to painstakingly construct an extensive administrative record that addressed the issue of congressional intent and the legislative history of the extent and manner of surface coal mining regulation in AVFs. In 1989, OSM adopted final definitions of agricultural activities and farming that were not challenged by either industry or environmental groups.

The definition of what constitutes farming on AVFs was not finalized for a decade after the first set of permanent program regulations were promulgated. During this time, guidance was uncertain for both OSM and the states, and, consequently, the issue of what would be included in areas where mining was to be prohibited was left unsettled.

One argument offered by proponents of a VER definition deferential to property rights is that the intent of Congress in including the VER proviso in SMCRA was to avoid takings of private property. Recently, the courts have held that SMCRA's prohibition of mining in AVFs could also be construed as takings. In a recent court case—Whitney Benefits, Inc. versus U.S. (926 F.2d 1169, Fed. Cir. 1991)—the U.S. Court of Claims and the

U.S. Court of Appeals held that applying the prohibitions to mining in AVFs constituted a taking, and they awarded more than $60 million to the owners of the coal rights. Including accrued interests and court determined legal costs, the total cost of the award may approach $150 million.

STATE PRIMACY UNDER SMCRA: CONCEPTUAL AMBIGUITY AND UNSTABLE CONSENSUS

A final policy legacy relates to the contours of the federal-state relationship under SMCRA. The legislative history shows that Congress wavered regarding which level of government was to have primary responsibility for regulation of surface coal mining. Initial versions of SMCRA assigned primary responsibility to the federal government, with the states in a supporting role. Section 101(f) of the act states that "because of the diversity in terrain, climate, biologic, chemical and other physical conditions . . . the primary responsibility for developing, authorizing, issuing and enforcing regulations for surface mining and reclamation operations should rest with the States." As one author observed, this is a paradox. The raison d'être for SMCRA was the failure of states to prevent environmental damage from surface coal mining. Yet, the congressional solution was to assign to the states the primary responsibility for implementing the remedial legislation (Rochow 1979).

Once a state program has been approved by the secretary of the interior and the state has in place a state law and regulations consistent with the federal regulations, the state is responsible for implementing the policy objectives of SMCRA. Under this scheme, coal operators must comply with state laws and regulations. At the same time, OSM maintains that it has authority to issue Notices of Violations (NOVs) in primacy states. This policy is a continual source of irritation to industry and to state regulatory authorities. OSM has issued regulations that provide some flexibility in the length of time OSM may allow states to respond to 10-day notices (TDN) issued by OSM as part of a federal inspection. In the preamble to the final rule that established these standards and the way the federal government is to carry out its oversight of states, OSM acknowledged that "what the Federal oversight should entail has been a continuing source of debate" (*Federal Register* 1988, 26731).

SMCRA continues to be the subject of heated debate and litigation concerning the extent to which OSM may also directly enforce regulatory requirements in primacy states. Under current regulations, OSM has created a dual system of enforcement where operators must comply with both OSM and state regulatory requirements. Congressional failure to state more clearly and specifically OSM's oversight and enforcement role continues to strain the federal-state relationship.

Even with final rule making on this matter, the issue is not fully settled. Industry has twice petitioned the secretary of the interior to initiate rule making on the issue of OSM's authority to issue NOVs during oversight. The secretary denied this aspect of both petitions, and the matter has been under litigation since 1989.

However, the legislative ambiguity present in SMCRA's regulatory scheme has some utility. It enabled Congress to point in two directions—to show deference to the states at the same time it made appropriate noises in the direction of federal responsibility and authority. This is not to say that, even in the best of situations, the federal-state relationship would be easy or free of conflict. In fact, a placid relationship could be evidence that one of the partners was not actively fulfilling its role. However, greater precision in legislative language would have minimized the endless rounds of rule making and consequent litigation that now seem to be required to resolve issues where Congress, for whatever reason, opted for ambiguity.

Lack of explicit statutory guidance has left both state and federal regulation with a policy gap that affects the smooth operation of cooperative federalism. Because this principal element in the SMCRA scheme of cooperative federalism was not well expressed in the act, OSM has had to cope with enduring problems in federal-state relations, as well as with troubling uncertainties for all those concerned with environmental protection and reclamation.

CONCLUSIONS

SMCRA was not a hastily enacted statute. It required years of concentrated effort, vigorous debate, and artful compromise to develop and to gain acceptance for many of its provisions. However, even after this extended consideration and revision, significant policy legacies remain that have retarded SMCRA's comprehensive implementation. Legislative ambiguity or avoidance of significant issues do not make the issues go away; rather, they lead to an extended process in which agency and interest groups resolve those issues or ambiguities in other forums. These arenas are no more expeditious, and in some cases no more conclusive, than the legislative process for resolving contentious or conceptually difficult issues.

REFERENCES

Bardach, Eugene. 1977. The Implementation Game, Cambridge Mass: MIT Press.

Congressional Record, House, 92nd Congress, 2nd Session, Oct. 11, 1972, pp. 35031, 35046, 35052.

Congressional Record, House, 93rd Congress, 2nd Session, July 25, 1974, p. 6705.

Congressional Record, House, 94th Congress, 1st Session, March 4, 1975, p. 5205.

Congressional Record, House, 94th Congress, 1st Session, March 18, 1975, p. 1908.

Congressional Record, House, 94th Congress, 1st Session, May 7, 1975, p. 13368.

Congressional Record, House, 94th Congress, 1st Session, June 10, 1975, p. 18008.

Congressional Record, House, 95th Congress, 1st Session, July 21, 1977, p. 7589.

Congressional Record, Senate, 1st Session, May 19, 1977, p. 8039.

Dunlap, Louise C. 1975. "An Analysis of the Legislative History of the Surface Mining Control and Reclamation Act of 1975." *Rocky Mountain Mineral Law Institute* 21: 29.

Federal Register. 1988 (July 14). Vol. 53 (135).

Legislative Proposals Concerning Surface Mining of Coal. 1971 (September 1). Committee Print, 92nd Congress, 1st session.

McElfish, James M., Jr. 1989–90. "Valid Existing Rights: Contingent and Equitable Rights—Not Real Property Rights (Another Attempt at a Definition)." *Journal of Mineral Law and Policy* 5(3): 624.

Moe, Terry M. 1989. "The Politics of Bureaucratic Structure." In John E. Chubb and Paul E. Peterson (eds.), *Can the Government Govern?* Washington, D.C.: The Brookings Institution.

Rochow, James. 1979. "The Far Side of Paradox: State Regulation of the Environmental Effects of Coal Mining." *West Virginia Law Review* 81: 559–93.

U.S. Congress, House. 1976 (August 31). Report of the Committee on Interior and Insular Affairs, to accompany H.R. 13950. Washington, D.C.: U.S. Government Printing Office.

U.S. Congress, Senate. 1971. Hearings Before the Senate Committee on Interior and Insular Affairs, U.S. Senate on S. 77, S. 630, S. 1160, S. 1240, S. 1498, S. 2455 and S. 2777. Pending Surface Mining Legislation November 16, 17, and December 2.

U.S. Congress, Senate. 1977. Hearings before the Subcommittee on Public Lands and Natural Resources, Committee on Energy and Natural Resources, Senate, February 7, March 1,2, and 3, 1977. Washington, D.C.: U.S. Government Printing Office.

3 Creating a Regulatory Agency: Profile of the Office of Surface Mining

Donald C. Menzel

Regulatory federalism is a term widely used to describe the American federal system. Its origins date back to the 1970s with the enactment of national legislation such as the Clean Air Act Amendments of 1970, the Federal Water Pollution Control Act of 1972, the Occupational Safety and Health Act of 1970, the Surface Mining Control and Reclamation Act of 1977 (SMCRA), and others. These ambitious laws formed a foundation for a complex, intergovernmental regulatory system that continues into the 1990s.

New regulatory laws often result in new bureaus and agencies that are expected to carry out the intent and will of the Congress. SMCRA, for example, created the U.S. Office of Surface Mining (OSM), an agency charged with the challenging task of striking "a balance between the protection of the environment . . . and the nation's need for coal as an essential source of energy" (Surface Mining Control and Reclamation Act of 1977, P.L. 95–87, 91 Stat. 445 1977). More specifically, the act called for the establishment of a "nationwide program to protect society and the environment from the adverse effects of surface mining operations" and the reclamation of mined areas left without adequate reclamation prior to 1977 that substantially degrade the quality of the environment or endanger the health or safety of the public. The law requires OSM, as the federal agency responsible for its implementation and oversight, to be both a regulatory and a public works agency.

THE BEGINNING

OSM was established in a White House Rose Garden ceremony on August 3, 1977, when President Jimmy Carter signed into law the Surface

Mining Control and Reclamation Act. This complex, controversial law was preceded by seven years of spirited debate and two presidential vetoes by Gerald Ford.

Secretary of the Interior Cecil D. Andrus selected Walter Heine (former chief of Pennsylvania's reclamation division) on December 16, 1977, to serve as the first director of OSM. Director Heine moved quickly to staff OSM and to establish a federal presence in the coal fields. By May 1, 1978, small staffs were assembled in regional offices at Charleston, West Virginia; Knoxville, Tennessee; Indianapolis, Indiana; Kansas City, Kansas; and Denver, Colorado. By the end of May 1978, an OSM inspector in Harlan County, Kentucky, issued the agency's first mine closure order.

Creating a new regulatory agency is an important first step in establishing an effective regulatory program. Personnel who believe strongly in its mission are often attracted to a new agency, and therefore they assign the highest priority to accomplishing the goals of the federal legislation.

Congressional framers of SMCRA seemed to have had this in mind when they created OSM and placed it in the U.S. Department of Interior. The Interior Department was a reasonable choice since it was the only federal agency that had experience in dealing with state coal surface mining programs, particularly in the western states where the federal government owns vast tracts of land. Still, the Department of Interior had limited experience that consisted primarily of conducting mining safety inspections under the Federal Coal Mine Health and Safety Act of 1969 and promulgating regulations in 1976 that covered coal mining operations on federal lands. Although it cannot be stated confidently that the decision to locate OSM in the Department of Interior rather than in the U.S. Environmental Protection Agency or the Department of Agriculture was a sound decision, OSM's placement did cause the agency some problems at the beginning. One high-ranking Carter-appointed OSM official put it this way:

> Locating OSM in Interior did have an impact on our mission. Let me give you an example. When we got ready to have the first hearings on the interim rules—formal hearings—I couldn't find anyone in the Department that had ever run that type of hearing or could tell you how to run one. So in a lot of respects, we were plowing new procedural grounds. It might have been easier if we had been in EPA where they were used to the type of rule making we were going through. (Interview with Paul Reeves, deputy director, Office of Surface Mining, Washington, D.C., September 24, 1981)

OSM's organizational home has been the subject of much congressional debate over the years. In 1988, the General Accounting Office (GAO), at the request of the chairman of the House Subcommittee on Environment, Energy, and Natural Resources, released a report indicating that coal industry representatives, environmental groups, and state mining officials believed it would not be wise to make OSM an independent agency or transfer its regulatory functions to another agency. They feared that such a move would further destabilize the agency and do little to improve the balance between environmental protection and resource development interests (U.S. General Accounting Office 1988, 5).

STRUCTURE AND OPERATION IN THE CARTER ADMINISTRATION

As federal agencies go, OSM was and is a small agency—fewer than one thousand full-time employees and, in 1980, a budget of nearly $180 million. In the Carter years, regulatory programs consumed $85 million while a larger sum, $95 million, was dedicated to abandoned mine reclamation programs. Moreover, most regulatory dollars and abandoned lands funds supported federal efforts and programs, not state programs.

OSM's creation in 1977 provided employment opportunities for many persons, especially state inspectors who were attracted to OSM by higher wages, job security, and the belief that working conditions would be better at the federal level. Additionally, some OSM staffers, especially in the Washington office, felt strongly about the agency's environmental mission and were ideologically in tune with the leadership (Hedge and Menzel 1985).

Decentralization was a guiding organizational principle under the Carter administration. Secretary Andrus and Director Heine believed that a decentralized OSM would allow the agency to be more flexible and responsive to state needs. The initial structure consisted of a headquarters staff in Washington, regional offices in five cities, and a set of district/field offices in each region. Functionally, each regional office mirrored the headquarter's office, with divisions consisting of management and budget, technical services and research, state and federal programs, inspection and enforcement, and abandoned mined lands. The regional directors reported to the director of OSM, not to the assistant directors in charge of each program area. At the end of the first full fiscal year of operation—fiscal year 1979—27 percent of OSM's staff (249 employees) were located in Washington, with all other employees working and living in other regions of the country (U.S. Office of Surface Mining 1979, 50).

This organizational arrangement, although rational and efficient looking on paper, became a problem as OSM leaders began to implement the law. For one thing, field-level activities—inspections—were initiated before regional offices were fully staffed and operational. As one regional director explained, "The zeal to get going caused OSM people to get inspectors in the field fast. We had inspectors in the field before regional directors were appointed, which was a disaster" (interview with Edgar A. Imoff, regional director, Office of Surface Mining, Indianapolis, Indiana, March 3, 1981). Personnel throughout the agency were learning their new roles and relationships and, not surprisingly, confusion often existed about who was in charge—Washington or the regional offices.

Matters were complicated further by the strong desire of OSM top managers to establish as quickly as possible a reputation for tough but fair-minded enforcement. OSM inspectors were instructed to "ticket" by the book—cite every violation of the interim program regulations. The Washington staff, not the regional offices, developed inspection and enforcement guidelines. Consequently, when problems of interpretation or practice developed in the field, they were usually taken to Washington for resolution. These organizational behaviors tended to undermine the authority of the regional offices, leading one regional director to assert that "we are so far down in the line management that our decisions are remade twice over" (interview with Donald Crane, regional director, Office of Surface Mining, Denver, March 27, 1981).

EARLY IMPLEMENTATION

OSM's initial strategy was to write elaborate rules and insist that the states follow them as closely as possible in designing their interim and permanent programs (see Shover, Clelland, and Lynxwiler 1986). Both "carrots" and "sticks" were used to help the process along. The carrots consisted of federal grants and, eventually, millions of dollars to reclaim mined out and abandoned lands. OSM, of course, had to develop regulations to administer the grant program. However, it was late 1980 before any extensive amounts of grant funds began to flow to the states. This situation caused many state officials to be critical of OSM. Not surprisingly, OSM officials viewed the grants as important vehicles for influencing the states. As one OSM official put it at that time, "We are blatantly coercive about state use of grants" (confidential interview, cited in Menzel 1981).

This attitude, in conjunction with an initial implementation strategy to bargain only when absolutely necessary, created considerable tension between OSM and state regulators. OSM's zeal to make SMCRA work at nearly any cost almost destroyed the agency in 1979 when state authorities became convinced that a congressional remedy was necessary. West Virginia's Governor Jay Rockefeller proposed that the Congress enact legislation to nullify OSM's rule-making authority and influence. The Rockefeller Amendment passed the U.S. Senate on September 11, 1979, by a vote of 68 to 26. It called for state rules and regulations to be consistent with the act, *not* with the *regulations* issued by OSM, as originally required. The Rockefeller Amendment never made it out of the House of Representatives where the father of SMCRA, Representative Morris K. Udall of Arizona, wielded considerable influence. Still, the mere fact that such legislation would be proposed and favorably acted on by one chamber of the Congress was enough to concern, if not frighten, agency leaders.

THE END OF THE BEGINNING

As 1980 approached, OSM managers quickened the pace of implementation. President Carter's sagging public popularity signaled the possibility that top OSM officials might not have a second chance to put SMCRA into place. This was also a period when dual regulatory programs were actually in place in most states, and permanent programs that were to be state directed and administered had yet to be approved by OSM. Federal inspectors were issuing citations and cessation orders in all major coal producing states, consistent with their authority to administer the interim regulatory program. At the same time, state regulatory officials were enforcing state programs. The result was that coal operators were regulated by two programs, two agencies, and two governments. Cooperative inspection and enforcement efforts did take place on some occasions, but for the most part each government went its own way.

Meanwhile, OSM leaders moved ahead forcefully and vigorously with their efforts to review state program proposals and award primacy to states that had developed programs consistent with OSM's rules and regulations and, therefore, with SMCRA. By the close of 1980, eleven coal-producing states had been granted partial or conditional approvals—Arkansas, Colorado, Indiana, Maryland, Missouri, Montana, New Mexico, North Dakota, Tennessee, Texas, and Wyoming. One month later, but before Ronald Reagan was sworn in as the 39th president, another five states—Iowa, Kansas, Oklahoma, Virginia, and West Virginia—were added to the approval list. The major coal-producing states of Illinois, Kentucky, Ohio, and Pennsylvania failed to gain primacy before the Carter administration left office.

THE "NEW" OSM

Ronald Reagan's presidential victory brought a dramatic shift in the federal government's approach to solving national problems, including the regulation of coal surface mining. Newly appointed Secretary of Interior James G. Watt and his OSM appointees took a different view of SMCRA and federal-state roles in implementing the law. Among other things, they believed that the implementation effort in the Carter administration had favored environmental protection over energy production, a stance that, in their judgment, was contrary to the act's intent to "strike a balance" between these twin needs. Additionally, they believed that SMCRA could be implemented more effectively by giving the states greater flexibility and responsibility to design programs that would fit under the federal regulatory umbrella. The newly nominated director, James R. Harris, articulated this attitude clearly in testimony before a congressional committee (Harris 1981). He stated that OSM should give the states the backing they need. "The only question at this point," he asserted is, "do you trust the states? Our answer is 'yes' " (Harris 1981).

Among the first tasks taken up by the newly installed leaders of OSM was to establish a management team to evaluate the agency's operating structure and put forward recommendations for reorganization. The prevailing view was that the regional structure created by the previous administration was not functional. As the then deputy director of OSM explained:

> My own practical experience with the regional structure was that in most every instance no decision was made in Charleston; it was made in Washington. The Regional Directors were put in decision making positions, but generally policy was made on a "catch-as-can" or a "catch-as-is" basis and it was always made in Washington. Moreover, even when a policy was made in a region it was reversed in Washington. (interview with J. Steven Griles, deputy director, Office of Surface Mining, Washington, D.C., October 15, 1981)

By April the details of the proposed reorganization were pulled together in a 70-page document; a month later the plan was made public through a press release issued by Secretary Watt. The reorganization plan, to be phased in over a 12-month period, would abolish the five regional offices, establish thirteen state liaison offices in the capital cities of the major coal producing states, create two technical centers (one in Pittsburgh and the other in Casper, Wyoming) and make changes in the Washington headquarters office. Washington operations were to be handled by three assistant directors rather than five who would report to the director. Secretary

Watt argued that the net result of these changes would be a savings of more than $2 million in administrative costs, a gradual reduction in personnel from a ceiling of 1,001 to 628 permanent, full-time employees (860 employees were actually working at the time of the announcement), and increased operating efficiency.

Secretary Watt's news release prompted a strong reaction from environmentalists and some members of Congress. Environmental groups accused the secretary of gutting the agency and intending to kill the regulatory program. Members of Congress whose districts contained OSM offices targeted for closure questioned the efficiencies Watt projected. Other critics charged that Watt had not consulted with key state, industry, and federal officials. A *New York Times* editorial described Secretary Watt as a "Washington princeling who ignores Congress and the public and simply does as he likes" (*New York Times* 1981).

Additional evidence that a "new" OSM was emerging can be found in the agency's budget for 1985. Those statistics show that federal funds flowing into state regulatory programs in the form of program grants increased from $21.7 million in 1980 to $36.7 million in 1985, a 69 percent jump. During the same period, OSM dollars dedicated to federal regulatory programs decreased by 4 percent. The agency spent $36.6 million in 1985 on federal regulatory programs in contrast to $38.0 million in 1980. Similar shifts can be seen in spending patterns on abandoned mine lands programs. In 1985, federal funds in support of state programs to reclaim abandoned mines was $257.7 million, which stands in sharp contrast to the $34.8 million spent by OSM for federal reclamation efforts.

REWRITING THE REGULATIONS

The new OSM that Secretary Watt envisioned was expected to fashion new rules—to repeal and rewrite those that, in his judgment, had contributed to a rigid and inflexible approach to implementing SMCRA. OSM leaders were accorded full presidential blessing for this effort through President Reagan's executive order of February 19, 1981. That order required federal agencies to review, revise, and, if necessary, eliminate all burdensome or counterproductive regulations.

OSM rule rewriters went to work quickly in 1981. Some 89 rule sections were targeted to be deleted, 329 sections were to be revised, 112 sections were to be combined with other sections, and 12 new sections were to be added. This effort consumed considerable staff time and energy in 1981 and 1982. It also alarmed environmentalists who quickly challenged the secretary's actions in the U.S. District Court in Washington (National

Wildlife Federation, et al., v. James G. Watt, et al., Civil Action No. 82–0320, April 16, 1982, U.S. District Court for the District of Columbia).

Perhaps the most significant rule change among the many proposed had to do with the Carter administration language that state program rules were to be "no less stringent" than federal rules. State officials had long taken issue with this language, arguing that it turned them into OSM clones and impeded their ability to fashion programs that would meet specific state needs. Sympathetic to this perspective, OSM managers put forward language that would require state rules and regulations to be "no less effective" than OSM's regulations. This new language essentially liberalized rule development and allowed states greater flexibility in designing their permanent regulatory programs.

CHANGING ENFORCEMENT STYLES

OSM leaders moved with equal vigor and aggressiveness to reorient the attitudes of the agency's inspectors, stressing the need for cooperation and persuasion rather than confrontation and coercion in dealing with coal operators and state regulatory authorities. A series of training sessions was held in 1982, intended in part to reshape the attitude of many OSM inspectors that they were still the primary regulators. Recognizing that this task might not be simple or straightforward, one OSM manager commented, "There's a possibility of some policeman-type mentality being maintained by the inspectors, but it's something we're aware of and we're going to change" (confidential interview cited in Menzel 1983). The agency's new slogan—credibility, cooperation, and compliance—was not to be taken lightly.

Another way to examine this shift in enforcement styles is to study OSM's enforcement activities during this period. Statistics show that in 1981 OSM conducted fewer inspections than it did in 1980, issued significantly fewer Notices of Violations (NOVs), and halted fewer mining operations (Menzel 1983). More specifically, in 1979 some 204 OSM inspectors conducted 20,306 inspections and issued 4,461 NOVs and 815 cessation orders (COs), numbers that contrast with 1981, when 156 inspectors conducted 27,070 inspections and issued 2,368 NOVs and 594 COs (Menzel 1983). In other words, in 1981 a smaller inspector corps made more inspections but issued fewer tickets than was the case in 1979. The ratio between the number of citations and the number of inspections decreased between 1979 and 1981. Part of this shift was due to the fact that OSM inspectors assumed a greater oversight role throughout this period. Still, there appears to be little doubt that OSM inspectors were doing things differently.

To gain credibility, respect, and cooperation from the states, OSM managers moved rapidly to grant primacy to all eligible states and to place the agency in an oversight role. By the end of fiscal year 1982, nine months into the President Reagan's first term, programs in seven states had been fully approved and programs in 17 other states had been conditionally approved. In the next year, 12 state programs were fully approved and 13 state programs were conditionally approved (Menzel 1983).

OSM's oversight program began to take form as early as March 1981 when guidelines were developed for drawing statistical samples to conduct inspections in states with approved programs. By the close of fiscal year 1983, OSM inspectors had conducted 4,912 inspections, issued 734 10-day notices to state regulatory authorities, written 77 notices of violation, and issued 27 cessation orders (U.S. Office of Surface Mining 1983).

Coal operators, state regulatory authorities, and environmentalists quickly disputed the vigor and accuracy with which OSM carried out its oversight duties. The National Wildlife Federation concluded that OSM's oversight program was ineffective and lacking throughout (National Wildlife Federation 1985). Indeed, the Federation attacked OSM, asserting that the agency was leaderless, adrift, and in a state of chaos (National Wildlife Federation 1985).

Between 1981 and 1984, under Secretary Watt, rules were changed and staff members were reeducated. Federal-state conflict, although not entirely absent, was reduced in frequency and intensity. Some conflict, however, continued, with much of it structured along classic interest-group lines. Environmentalists were relentless in monitoring and challenging OSM's actions, both in the courts and in the Congress. Principal court victories came in 1984 and 1985 when a federal judge ruled that OSM must rewrite previously rewritten rules dealing with 79 issues (In re Permanent Surface Mining Regulation Litigation 11, No. 79–144, U.S. District Court for the District of Columbia; the actual decisions consisted of three rounds of judgment and order memorandum issued on July 6, 1984, October 1, 1984, and July 15, 1985).

OSM BECOMES OSMRE

James R. Harris resigned in March 1984 and was succeeded by several acting directors—including his eventual replacement, Jed D. Christensen. Christensen became acting director in April 1985, but his nomination was not confirmed by the Senate until May 1986. Management instability was also caused by the turnover of three deputy directors between late 1983 and 1985. One of those officials, Steven Griles, moved to an assistant secretary post in the Department of Interior.

Christensen's appointment was widely viewed as an effort by Secretary of Interior Donald Hodel (Watt resigned in 1983) to infuse a healthy dose of professional management in OSM. Christensen had no previous knowledge or experience in coal surface mining regulation before he was appointed acting director. He joined the Department of Interior in 1981 and served as the deputy assistant secretary for water and science before moving to OSM. Before his appointment at the Department of Interior, he had served as the chief financial officer of several municipalities in western states.

Prodded by a continuing stream of criticism, OSM leaders, under the direction of Secretary Hodel's office, drafted a blueprint to manage the agency and carry out the dictates of the law. The *Management Action Plan* was released in draft in November 1985. This detailed, 177-page document required three months of intensive work to complete and was described as "a road map for change" that would lead to OSM's renaissance.

The document focused on six elements that, collectively, defined the agency's mission: (1) assure the reclamation of mined coal lands; (2) maintain a viable coal mining industry; (3) provide a stable, consistent regulatory and policy framework; (4) encourage state authority and responsibility; (5) utilize an open public process; and (6) provide a foundation of research and technology development to reduce coal mining's adverse impacts.

The political rhetoric of the Harris regime—credibility, cooperation, and compliance—was to give way to the managerial rhetoric of consistency, certainty, and consensus. However, no one—not environmentalists, nor coal lobbyists, nor state regulatory authorities, nor political leaders—embraced the plan. Rather they were highly critical. As a result, more than 2,000 draft copies of the management action plan that were printed for circulation were destroyed; only two copies were kept.

CHRISTENSEN'S AGENDA

Christensen believed that managing meant more than producing results; it also meant image. Among the first things he did was to rename the agency. OSM became OSMRE—the Office of Surface Mining Reclamation and Enforcement. Christensen also lent his support to a series of meetings to bring together officials from federal and state regulatory authorities, to exchange views and share information. Meetings were held in cities in coal-producing states and in cities such as Reno, Nevada, and Tampa, Florida. The meetings were well attended and conducted within the spirit of regulatory negotiation.

Christensen's initial internal management agenda consisted of three priorities: rule making, information management, and delinquent penalty collection. When Christensen arrived at OSM, rule making was bogged down; nearly 80 final rules had been challenged in the U.S. District Court either by the coal industry or by environmental groups. Consequently, he worked to get rules finalized in a timely manner. Next he faced a major problem of information management. Millions of dollars had been spent on computer systems "with absolutely no coordination or game plan on what they were to do." Finally, OSM had more than $200 million in uncollected civil penalties that dated back to 1978. This situation had generated considerable criticism from Congress and the inspector general. Christensen organized a special group "to bring order to the process and collect or eliminate the debt." While Christensen believed this effort was successful (letter from Jed D. Christensen, July 9, 1990), others felt that writing off millions of dollars of uncollectible fines to deal with this problem, which was part of the approach, could hardly be defined as successful.

Despite his management priorities and intentions, Christensen was unable to shield OSMRE from harsh criticism, and morale within the agency continued to sag. Consequently, in 1987 he formed a 13-member human resources task force to generate ideas to improve the morale of agency personnel. The task force, comprising a cross section of agency staff members, including secretaries and middle-management personnel from state OSMRE offices, recommended that a management consultant firm be hired to conduct an organizational climate survey. Seventy-two percent of the agency's personnel responded to the survey conducted in October 1987. The survey results reported in February 1988 in a document titled *Bridge to Excellence* revealed that: (1) employees believe favoritism is practiced in administering hiring, promotions, and awards programs; (2) employees believe poor communication exists throughout OSMRE; (3) employees believe the agency is micro-managed; (4) employees want more responsibility for work assignments; and (5) many employees do not have a clear sense of the agency's mission.

The survey also found that many OSM staffers believed the agency was dominated by politics, and political motivations were behind most decisions. Christensen took a different view:

To me it's very naive for people to think an agency like OSM isn't going to have some politics involved in it. When you have diverse groups of people interested in the agency—the environmental community which is very political, industry people who are very active, Congress who's scrutinizing you like crazy, and an administration who has a certain philosophy, you've got to recognize that there's politics involved in almost everything. I feel people were naive not to recognize that there were going to be some political directions in the kinds of things we did.

However, I don't agree that the agency was any more politically motivated than any other agency I saw in Interior. They were all led by politically appointed people who had an agenda. We were trying to pull off our agenda, and to say that it wasn't a politically motivated agenda is ridiculous. We had some things we wanted to accomplish and followed the philosophy of the administration, so it doesn't bother me that some things we did were politically motivated in the sense that we tried to run an agency fairly but remembering that this was an administration that did not like regulation. The Reagan administration was caught administering a regulatory agency with an extreme environmental law and trying to do it in a fair way. It was a tough spot to be in but I think we did it as fairly as we could. (Correspondence, September 29, 1988)

In March 1988, Christensen resigned as director of OSMRE. His management style, in effect between 1985 and 1988, was redirected when he departed, and in anticipation of the 1988 presidential election.

Christensen was replaced by Robert Gentile, who had joined the Department of Interior in April 1987 as liaison for coal affairs. Gentile had been recruited from the ranks of the coal industry, having owned and operated a coal mining firm in eastern Ohio. As liaison for coal affairs, Gentile reported directly to Griles.

Griles and Gentile shared similar views of OSMRE and the future of the agency. They believed the agency should work with the states to the fullest extent possible in order to make SMCRA successful. In mid-1988, Griles asserted that the next two to three years would determine the success of SMCRA (interview, July 6, 1988, in Washington, D.C.). He believed that if OSMRE took a hard line toward the states, there could be massive abandonment of the program by the states. Gentile said he would be pushing for a "stable, understandable approach" in carrying out the act (interview, July 6, 1988, in Washington, D.C.). In his words, "we're here to help the states, and believe that a federal-state partnership is workable."

The seemingly transient nature of the directorship has not ended. In August 1989, Gentile was replaced by Harry Synder of Kentucky who, in turn, has undertaken yet another major reorganization of OSM. Moreover, the agency remains vulnerable to political firestorms.

A General Accounting Office study released in February 1989 reported both progress and lack of progress in the management of OSMRE (U.S. General Accounting Office 1989, 6). The study, based on interviews with key participants in the summer of 1988, found that agency leaders had implemented more than 40 of the 65 recommendations set forth in 1985 in a study called the Shafer Management Report. However, several key recommendations had not been acted upon. One of these, for example, was:

> The Department and [OSMRE] need to make clear [OSMRE's] role and how SMCRA is to be implemented. The Secretary needs to communicate his intent to comply fully with the Act to [OSMRE] employees, industry, environmentalists, and other affected parties. The Secretary should also make clear that the Act will be implemented consistently and that in-house conflicts will be resolved fairly and expeditiously. (U.S. General Accounting Office 1989, 6)

General Accounting Office investigators could find no written evidence (memos, for example) to show that the secretary of interior intended to comply fully with the act or to resolve conflicts fairly and expeditiously. Neither did they find any evidence that OSMRE officials had conducted a thorough review of the agency's working relationships with external organizations (another recommendation), particularly the relationship between OSMRE and the solicitor's office. Finally, OSMRE had taken no action to develop a system that would identify and track issues relevant to OSMRE and SMCRA. Such a system would track the status of the issue, identify responsible implementation parties, estimate needed resources, and monitor completion dates and reporting requirements (U.S. General Accounting Office 1989, 10).

CONCLUSION

The creation of OSMRE as an effective regulatory agency continues. This chapter has described and documented the long and often difficult plight of a regulatory agency that few like and many criticize. Neither feared nor loved, OSMRE may be a victim of failed federalism, failed management, failed politics, or failed policy—or perhaps all of those.

Failed federalism is a distinct possibility. A national regulatory program that depends heavily on the states to succeed may be no more likely to succeed than a national regulatory program that depends heavily on and is centrally directed by Washington. The partial preemption regulatory approach that characterizes SMCRA and that has been the framework for other federal programs does not seem to have worked as well for coal surface mining regulation. The states and the coal industry have often allied themselves, either willingly or unwillingly, over the past 15 years, in an effort to stymie OSMRE. Coal politics remains a powerful influence in Washington.

Failed management cannot be ruled out as a possible explanation for the plight of OSMRE. Still, that explanation alone does not seem sufficient. No one has presented evidence to suggest that OSMRE managers are less well trained or less dedicated than other federal civil servants.

Failed politics might be a third explanation for the performance of OSMRE. If politics is indeed the art of the possible, with differences tempered by compromise, then there is plenty of evidence that SMCRA actors—federal and state authorities, environmentalists, and coal industry representatives—opted out from a political success story many years ago. The question might well be: "What might be done to create a regulatory process that encourages reasonable and fair-minded behaviors by the principal antagonists?"

Finally, it is not unreasonable, given the history of this policy, to conclude that the basic design of P.L. 95–87 is the culprit. Perhaps only a sweeping revision of the law can rescue the policy and OSM from the regulatory quagmire that seems to be their destiny.

It is difficult, perhaps impossible, to disentangle OSMRE's tormented history from the actual implementation of the law. This is partly due to the fact that even in the 1990s there is little objective data that the principal actors can draw upon to determine the "success" or "failure" of the law. Impressions, of course, abound. However, there appears to be little consensus among environmentalists, coal operators, or regulatory authorities—state and federal—about how things seem to have turned out after nearly 15 years of implementing P.L. 95–87.

As far as the future of OSMRE is concerned, the only thing that can be said with confidence is that it is likely to be an uncertain future. It seems clear that the failures of the past have served few interests, and every effort should be made in the decade ahead to strike a working balance among federalism, management, and politics. Environmental protection and abundant energy supplies remain important goals of the American people.

REFERENCES

Harris, James R. 1981 (August 5). Statement Before the House Interior and Insular Affairs Committee. Mimeograph.

Hedge, David M., and Donald C. Menzel. 1985. "Loosening the Regulatory Ratchet: A Grassroots View of Environmental Deregulation." *Policy Studies Journal* 13: 599–696.

Menzel, Donald C. 1983. "Redirecting the Implementation of a Law: The Reagan Administration and Coal Surface Mining Regulation." *Public Administration Review* 43: 411–20.

———. 1981. "Implementation of the Federal Surface Mining Control and Reclamation Act of 1977." *Public Administration Review* 41: 212–19.

National Wildlife Federation. 1985. *Failed Oversight*. Washington, D.C.

New York Times. 1981 (June 18).

Shover, Neal, Donald A. Clelland, and John Lynxwiler. 1986. *Enforcement or Negotiation: Constructing a Regulatory Bureaucracy*. Albany, N.Y.: State University of New York Press.

U.S. Department of the Interior. 1985 (June). *Management Review of the Office of Surface Mining*. Washington, D.C.: Office of Surface Mining.

U.S. General Accounting Office. 1989 (February). *Office of Surface Mining Response to Management Review Recommendations*. Fact Sheet for the Chairman, Committee on Interior and Insular Affairs, U.S. House of Representatives. Washington, D.C.: U.S. Government Printing Office.

———. 1988 (June). *Surface Mining: Transferring Interior's Surface Mining Regulatory Function*. Report to the Chairman, Environment, Energy, and Natural Resources Subcommittee, Committee on Government Operations, U.S. House of Representatives.

U.S. Office of Surface Mining. 1983. *Annual Report 1983*. Washington, D.C.: Office of Surface Mining.

———. 1979. *Annual Report 1979*. Washington, D.C.: Office of Surface Mining.

II Interest Groups and Implementation

4 The Surface Mining Control and Reclamation Act and Changes in the Coal Industry Structure

Harold P. Quinn, Jr.

Perhaps no environmental legislation has confronted a single industry with such a comprehensive regulatory approach to the way of doing business as has the Surface Mining Control and Reclamation Act of 1977, P.L. 95–87 (SMCRA). Unlike environmental legislation of the 1970s directed at the impact of many industries upon one natural resource, SMCRA focuses upon one industry and its effect upon several resources. Coal companies find that all aspects of development and production are influenced by SMCRA's prescriptions that cover everything from coal recovery to land management practices.

The preenactment history of SMCRA is replete with prognostications that a comprehensive law with detailed prescriptions for mining and reclamation would spell the demise of the small operator. To the casual observer, the substantial exit of small operators from the coal industry appears to validate that prediction. However, more fundamental forces are at work that better explain the causes for this evolution in industry structure. Certainly, SMCRA has accentuated the effect of these market forces, and, in some instances, has accelerated their inevitable consequences. While SMCRA has not changed the fundamental considerations for coal resource development, it has certainly influenced them.

To develop responsive and relevant policy, those responsible for implementing SMCRA must understand the fundamental forces that cause changes in coal industry structure and practice. SMCRA policy makers must remember that the quest to remain competitive in today's energy markets is a principal consideration in the resource development and production equation. Moreover, this equation is influenced not only by

SMCRA, but also by the numerous, and sometimes duplicative, laws that pertain to coal mining and coal markets.

SMCRA's success will depend largely upon the capability of those who administer the law to recognize and to understand the fundamental changes in the industry and to formulate policies that are responsive and relevant to the evolving industry. If the regulatory program does not evolve with the industry and its experience under SMCRA, uncertainty and conflict will increase; these, in turn, will aggravate political pressure for legislative reform.

CHANGES IN INDUSTRY STRUCTURE

Policy makers must discern the changes that occur in the coal industry. They must understand the reasons for those changes and recognize how they affect the regulatory program. Like changes in any industry, those in the coal industry are driven primarily by economics. Thus, alterations in industry structure and practice result from the quest to remain competitive in today's fuel markets. Gains in productivity supply the key to remaining competitive. The public derives two benefits from the coal industry's quest for competitiveness, or gains in productivity: (1) a competitively priced product, generally in the form of lower energy costs, and (2) more efficient and higher quality land restoration. The coal industry's experience demonstrates that mine productivity and sound land restoration practices not only are compatible, but are essential for an efficient operation.

A look at several indicia over a ten-year period readily discloses that industry structure has altered significantly since SMCRA was enacted. The number of so-called major coal producers, those producing in excess of three million tons annually, has increased, as has their overall share of production. In 1976, 34 companies accounted for 57 percent of all coal produced. In 1986, 52 firms accounted for 75 percent of all coal production. This change in the number of major producers, along with the increase of their market share to three-fourths of all production, has come at the obvious expense of smaller firms.

Who are these major companies? Fewer are the so-called independents, closely held companies, and more are the integrated energy companies, or utilities engaged in backward integration, particularly in the western United States.

During this same 10-year period, the number of producing mines fell 32 percent, from 6,161 to 4,203. At the same time, production jumped 30 percent, from 685 million tons a year to 886 million tons. This trend continued with fewer than thirty-nine hundred mines producing more than

946 million tons of coal in 1988. More production from fewer mines further confirms the trend of the major producers increasing their share of domestic production. In fact, the major firms that operate larger mines account for most of the increased production. The number of mines that produce less than one hundred thousand tons annually (the level that SMCRA originally designated as the threshold for its Small Operators Assistance Program) decreased 65 percent, from 5,100 to 1,827, between 1975 and 1987. The number of mines that produce more than one hundred thousand tons increased from 1,074 to 1,211, with the largest gain shown for those mines that produce in excess of one-half million tons of coal annually.

Just as SMCRA embodies a pluralistic approach to achieve its policy objectives because of regional diversity, so must policy makers recognize industry changes on a regional basis as they implement the regulatory program. Most of the restructuring of the coal industry occurred in the Appalachian coal region (see Figure 4.1). Most of the industry growth, in terms of production and significant market entry, occurred in the western region. Overall, the interior region remained relatively stable in the number of producers and the number of mines, as well as in overall production levels; however, significant geographical shifts occurred within this region for both new production and market entry.

Coal production grew rapidly in the western region between 1976 and 1988, increasing from 111 million tons to 307 million tons annually (see Figure 4.2). This growth was accompanied by the entry of oil and gas companies into the coal industry, with seven oil companies among the top 20 producers in the region. The western region includes fewer firms and fewer mines than other regions; the mines tend to be larger-scale surface mines working in favorable mining conditions, with coal seams often ranging between 40 and 70 feet in thickness. Stable market conditions have been associated with the western region, with most of the coal produced under long-term sales arrangements with utilities. However, coal ownership patterns differ significantly from those of the two other regions. Most of the coal is controlled by the federal government, with the remaining holdings of significance located on railroad lands and Indian lands. In 1989, production on federal lands accounted for 68 percent of all coal production in the western region.

Most of the interior region's coal production occurs in two separate areas: in Illinois, Indiana, and Western Kentucky, and in the lignite fields of Texas. While coal production in the interior region as a whole increased, most of the growth occurred in the production of lignite in Texas. Twelve of the 20 largest producers entered the market since the passage of

Figure 4.1
U.S. Coal Mines by Region, Number (1,000), and Percent Share

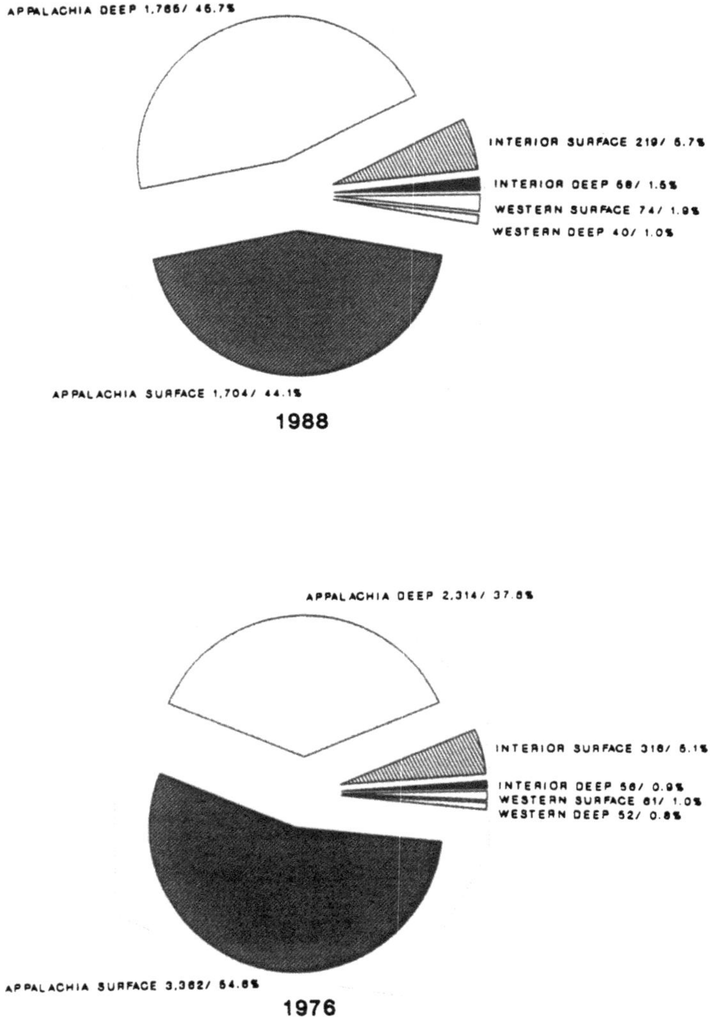

SOURCE:DOE/EIA/NCA

Figure 4.2
U.S. Coal Production by Region, Tons (millions), and Percent Share

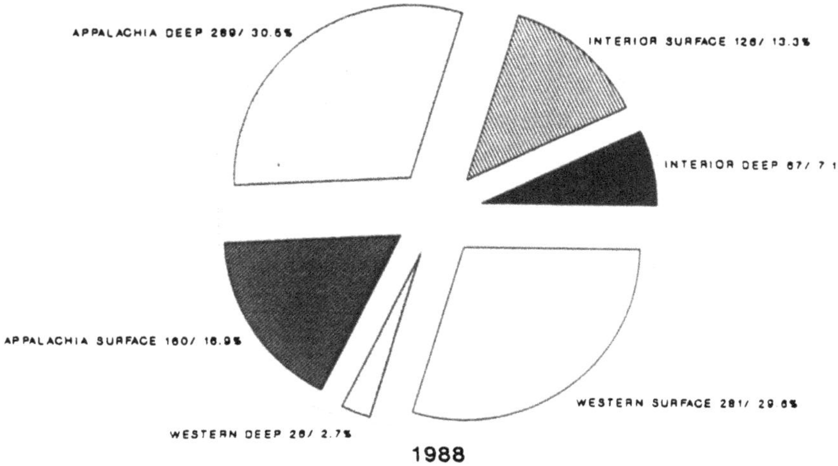

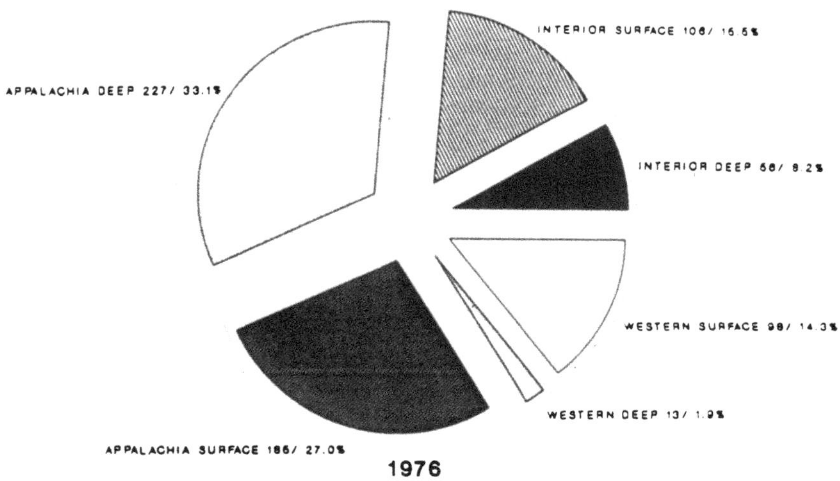

SOURCE:DOE/EIA

SMCRA, and most of them operate large lignite mines in Texas to supply mine-mouth power plants. At the same time, the number of firms that produce more than one million tons annually grew from 17 to 24; however, the total market share of the 20 largest firms fell from about 82 percent to about 75 percent of the region's production. It would appear that smaller firms were able to increase their share of production, primarily in Indiana. The stagnant production in the upper interior region follows the effects of the Clean Air Act; existing utilities were forced to use lower-sulfur coal to avoid the expenses associated with installing stack gas scrubbers. In the future, the interior region will face similar constraints as well as increased competition, particularly from Appalachian and western low-sulfur coal used by utilities to comply with the revisions to the Clean Air Act enacted by Congress in 1990.

The Appalachian region remains the most diverse region in terms of the types of coal, markets, methods of production, size of operations, and coal producer profile. While production jumped from 412 million tons in 1976 to more than 444 million tons in 1980, it remained relatively flat through most of the 1980s, with increases in West Virginia, Virginia, and eastern Kentucky, and declines in Pennsylvania, Ohio, Tennessee, and Alabama.

The significant change relates to who is mining the coal. Both the number and the size of firms have changed since SMCRA. Appalachia accounts for most of the departure of the small producers. Most states in the Appalachian region have about one-half the number of mines today that they had in 1975. In some areas such as Alabama and eastern Kentucky, the number of mines has declined to one-third from the boom times in the mid-1970s. Small producers mine 50 percent less coal than they did in the early 1970s, and the 20 largest firms produce more than half of all production in Appalachia. Nonetheless, Appalachia still accounts for 90 percent of all mines in the nation (see Figure 4.1). Moreover, since 1978, underground production has increased significantly, reversing the trend of earlier years (see Figure 4.3). Whereas before, surface mines comprised almost 60 percent of all mines in Appalachia, today underground mines account for more than 50 percent of all mines and almost two-thirds of the production in Appalachia.

Trends in coal markets generally explain the reasons for the changes in industry structure. To varying degrees, interfuel competition in energy markets and competition within the coal industry brought a deteriorating price structure to many coal markets. A deteriorating price structure for the product caused smaller, marginally efficient mines to close and brought about the demise of the smaller producers. The price squeeze affected primarily smaller producers who relied heavily upon the spot market for

Figure 4.3
Production by Region

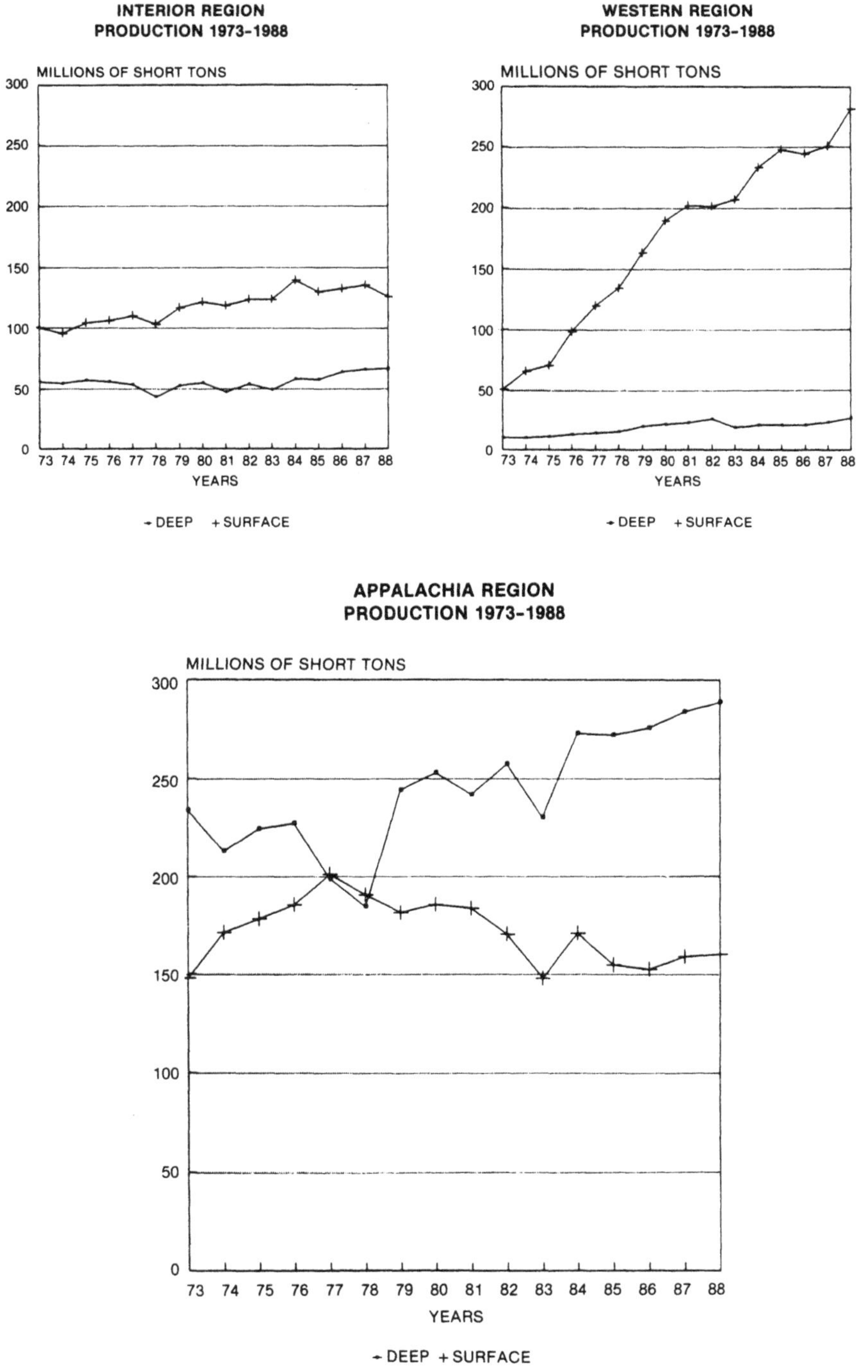

SOURCE: EIA/DOE

Figure 4.4
U.S. Coal Mining Labor Productivity Underground, Surface, and Average, 1970–1988

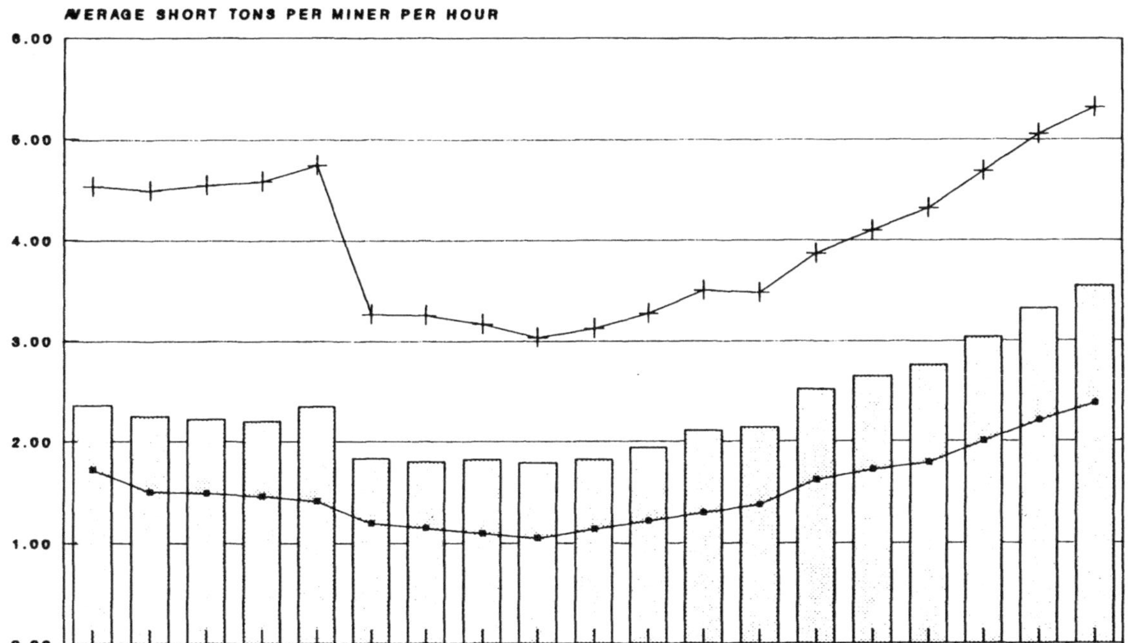

	1970	1971	1972	1973	1974	1975	1976	1977	1978	1979	1980	1981	1982	1983	1984	1985	1986	1987	1988
UNDERGROUND	1.72	1.50	1.49	1.46	1.41	1.19	1.14	1.09	1.04	1.13	1.21	1.29	1.37	1.62	1.72	1.79	2.00	2.21	2.38
SURFACE	4.53	4.49	4.54	4.58	4.74	3.26	3.25	3.16	3.03	3.12	3.27	3.50	3.48	3.87	4.10	4.32	4.69	5.06	5.32
AVERAGE	2.36	2.26	2.22	2.20	2.35	1.83	1.80	1.82	1.79	1.82	1.94	2.11	2.14	2.52	2.65	2.76	3.04	3.32	3.55

YEAR

UNDERGROUND — SURFACE — AVERAGE

SOURCE EIA/DOE

their sales. The average price of coal declined 10 dollars a ton in real terms in just 10 years (1975–1985). In the debate over the impact of SMCRA legislation economic analysts forecast that coal prices would exceed 50 dollars a ton. Those forecasts have proven well off the mark, with the average price of coal about 22 dollars a ton in 1988—the sixth consecutive year of decline.

To survive within a deteriorating price structure, coal producers must decrease costs and boost productivity. With tight margins between costs and price, producers will spread fixed costs over more units of production, with larger operations using newer technology. The result is that fewer companies produce more coal from fewer mines. Between 1976 and 1988, productivity rose from 1.80 to 3.55 short tons per miner hour, and the average production per mine doubled (see Figure 4.4). Much of the boost in productivity came from the large western surface mines, although in Appalachia the trend has also been toward larger surface mines that use in steep terrain the types of equipment deployed in area mines in the Midwest. Underground mining productivity rose with the use of longwall mining systems and more efficient use of continuous miners.

Merger activity in the 1980s also contributed to the trend toward fewer and larger coal firms. Several factors generally provided the impetus for that merger activity, including: (1) acquisition of a reserve base that enhanced a company's strategic market position; (2) diversification of the reserve base into low-sulfur or "compliance" coal; and (3) to some extent, acquisition of firms with favorable long-term contracts that could be served from the acquired or existing capacity. The decision of some firms and some industries, particularly steel, to leave the coal industry prompted some of the acquisition activity.

In sum, market forces comprise the principal reason for the changing structure of the coal industry. To survive in the face of a deteriorating price structure, coal producers were forced to boost productivity. Marginal mines and small producers became the casualties of that price squeeze. In other words, the industry structure changed as it became more efficient.

SMCRA'S EFFECT UPON THE COAL INDUSTRY

For some in the coal industry, the demands of SMCRA hastened the inevitable consequences that follow from market forces. A few of the casualties were, perhaps, not up to the task prescribed by the law. The remaining casualties, the smaller producers, were as capable as any of adjusting to SMCRA from an operations standpoint, but they simply were at a competitive disadvantage compared to the larger, more productive mines. The efficiencies at larger operations extend beyond direct operating costs, to compliance and reclamation. For many, the additional costs associated with SMCRA exacerbated the fundamental market forces that brought tighter operating margins upon everyone. As a general matter, small producers possessed less staying power to weather the continual downturn in price, despite their efforts to improve productivity.

SMCRA has not dramatically altered the fundamental considerations for evaluating coal mine development, but it has influenced them. For coal producers, both development and operating considerations are influenced by SMCRA's prescriptions that cover everything from coal recovery to land management practices. Accordingly, meaningful evaluations of mineable reserves, operating costs, and development schedules must include assessments of the protection measures found in SMCRA.

SMCRA has left the coal producer no choice but to integrate the planning of mine development with the management of other natural resources, both during and after mining. In short, a coal company must become equally proficient in the skills of land management and coal mining. This adjustment has been more painful for some than for others in the coal industry. Many of the principles embodied in SMCRA existed previously in state laws. In fact, SMCRA borrowed heavily from existing state laws. However, the federal law departs from many of its state predecessors in its substantive and procedural detail.

The early effects of SMCRA upon industry performance came during a period when the coal industry was still adjusting to a myriad of federal programs that related to the environment and to the safety of miners. While coal production grew to meet the increased demand of the 1970s, productivity declined as companies changed practices to meet the rigors of new regulatory programs. Once the industry perfected the methods and technology to meet these new demands, it experienced an increase in operational efficiencies that continues today.

Only future performance will disclose whether current operating strategies can continue to reduce the basic costs of producing coal. A question still remains whether SMCRA policy development will accommodate new and innovative approaches to mining and reclamation that boost efficiency.

IMPLICATIONS OF A RESTRUCTURED INDUSTRY TO THE FUTURE OF SMCRA

The changes in the coal industry structure, with a few exceptions, do not hold significant implications for the efficacy of SMCRA itself. These changes do, however, substantially affect policy development. The ability of policy makers to understand the changes in the industry and then to develop policy relevant, or responsive, to these changes will largely determine the future effectiveness of the regulatory program under SMCRA. Again, in the absence of policy development that accounts for changes in both the industry and future approaches to mining and reclamation, pressure will emerge to substitute corrective legislation.

Examination of several components of SMCRA's regulatory regime illustrates some of the considerations an evolving industry should present for the administration of the program by federal and state governments alike.

Resource Demands and Allocation

In recent years, both federal and state agencies have experienced fiscal constraints that affect their ability to augment their programs. Accordingly, program managers must rely upon their skills to allocate existing resources to meet shifting program demands. Under SMCRA, permitting and enforcement comprise the unit operations of the program and require the most resources in terms of personnel and costs. However, continual changes to the federal rules continue to exert unexpected demands for resources to serve the auxiliary functions of preparing and reviewing basic regulatory standards at both the federal and state levels.

The changes in industry structure should alleviate the demands for some program resources. With fewer mines, demands upon the permit and enforcement functions decline. Frankly, the market place has made future enforcement easier. Not only do the regulatory agencies have fewer mines to inspect, more of the mines are operated by larger and better capitalized firms.

The enforcement function has also become easier as the operators with the riskiest profiles have exited from the market. This is not to say that small operators, per se, comprise a noncompliant group. In fact, many small operators carry out the highest quality and the most innovative reclamation. However, the fact remains that small operators with high exposure to spot market trends often suffer immediately from the strains attendant to a downturn or a shift in markets. And, unfortunately, the market does not discriminate between the operator who has a reputable compliance history and the operator who is recalcitrant. Some may suggest that a rational enforcement policy should embody such a distinction.

Despite the changes in industry structure, enforcement will not become easier if policy makers remain preoccupied with the past. The so-called Applicant Violator System (AVS) reflects, in part, policy development that ignores the reality of the industry evolution. The premise behind the AVS is to identify recalcitrant operators—those from the initial regulatory program of the late 1970s and early 1980s—and ban them from continued participation in the permanent regulatory program. AVS implements this objective by screening all new permit applicants and flagging those with uncorrected violations. This task appears simple enough. However, after five years of experience, critics of the Office of Surface Mining have summarily concluded that the system is a failure, simply because an insufficient number of recalcitrant operators have been identified and denied new permits. This indictment of AVS charges that the data base and the computer program are inadequate.

Surprisingly, no one has bothered to examine the system's underlying premise in light of industry trends. The change in industry structure generally has resulted in the recalcitrant operator—the target of AVS—leaving the industry. The barriers to reentry for those operators include their inability to raise the substantial capital costs necessary to operate a mine large enough to compete in today's market. In short, AVS represents, at best, a passive enforcement tactic that cannot preclude what largely will not occur anyway—recalcitrant operators from the past seeking permits in the future.

Another reason AVS has not met what some view as unrealistic expectations is that efforts and resources have been diffused to implement a national system for a regional issue. With few exceptions, AVS has remained irrelevant to the type of industry and the enforcement issues encountered in the western and interior regions. Few, if any, thinly capitalized operators who left problems in Appalachia have the capability or desire to relocate to the interior or western coal-producing regions. Ironically, as the industry structure evolves in Appalachia, AVS will become increasingly irrelevant to that region. At this point, AVS provides little in the way of deterrence, while it inflicts an enormous burden on that segment of the industry the policy makers claim is not the target. A more focused and aggressive use of the considerable resources expended on AVS may have yielded more desirable results.

As fiscal pressures increase, policy makers should not discount the significance of changes in the industry and how those changes might affect the allocation of limited resources.

SMCRA's Performance Standards

Changes in the coal industry structure do not appear to significantly affect the substantive requirements of SMCRA. The statutory objectives are firmly grounded in elementary principles of resources management. Nonetheless, policy makers must remain mindful of and open to evolving mining practices to meet these objectives more efficiently.

The first attempt to flesh out SMCRA's prescriptions for mining and reclamation resulted in unyielding design standards. The Department of the Interior's regulatory reform efforts in the early 1980s gave the program much-needed flexibility through performance-based standards. The switch from design standards to performance standards accommodates regional differences in mining conditions, land use, and industry structure and allows each state the opportunity to fashion regulations that meet its particular needs. While a design approach to regulation stymies innovation, a performance-based approach can better accommodate new technology and advancement in mining and reclamation practices. Like any sound strategy, performance goals keep things simple and systematic, with minimal changes to accommodate new tactics. In sum, the inherent flexibility attendant to performance standards remains more responsive to an evolving industry.

Changes in the industry will often provide the occasion to reevaluate policy under SMCRA. Reclamation bonds come to mind as an issue that deserves evaluation in view of the trends in the coal industry. SMCRA requires that each coal operator post a bond sufficient to ensure the performance of reclamation. The amount and form of the performance bond will vary, depending upon the mine, the selected postmining land use, and the requirements of the state regulatory program.

The framers of SMCRA assumed that a surety industry existed with sufficient capacity and desire to service the coal industry; that assumption now appears to be severely mistaken. Before SMCRA, coal industry bonding requirements were substantially different from those embodied in SMCRA. Traditionally, both the amount and the extent of the reclamation obligations the surety industry underwrote shared much in common with their major line of business, such as construction projects. Before SMCRA, most states established generally flat rates for bond amounts, and the release of the bond coincided with the completion of reclamation activities. SMCRA introduced a more complex scheme, with bond amounts calculated to cover the additional regulatory costs attendant to SMCRA and a prolonged period of responsibility before final release—five years after the completion of reclamation in the East, and ten years in the West. Although the statute allows for a phased reduction of the bond amount, concurrent with the completion of certain reclamation activities, the maximum allowable reduction for each phase bears no resemblance to the actual costs of those activities. For instance, SMCRA allows the release of 60 percent of the bond amount upon the completion of backfilling and grading of a mined area. The actual cost of this activity generally exceeds 80 percent of the total costs for all reclamation.

In retrospect, it is not surprising that the industry has begun to experience increasing difficulty in securing surety services to satisfy the performance bond requirements. The substantial change in the type and extent of the obligations underwritten has caused many surety companies to eschew this line of business. Sureties can obtain better returns in other types of business without tying up their capital for the prolonged periods associated with SMCRA bonds. The nexus between SMCRA and shrinking surety capacity for reclamation bonds appears undeniable. Studies by government and industry show that the drop in the number of surety companies providing the service coincides with the implementation of SMCRA's bonding provisions through state programs.

What some have called the bonding crisis has affected many segments of the coal industry. But the small operator has felt the greatest shock from diminished capacity in the surety industry. The changes SMCRA brought to the bonding function were further compounded by the price squeeze experienced in coal markets. Sureties evaluate candidates for reclamation bonds in the same manner as they evaluate any credit risk. Accordingly, a company's financial position and market stability comprise the principal considerations in underwriting reclamation obligations. Even though a company might have an excellent reclamation history, that factor will not usually override financial considerations. Thus, one can quickly surmise why smaller operators whose cash flow depends largely upon a volatile spot market have borne the initial brunt of the bond crisis.

Alternative bonding vehicles could alleviate some pent-up pressures from the unfulfilled demand for surety coverage. Certainly, SMCRA provides considerable opportunity to develop alternative vehicles to meet this statutory requirement. Yet, many of the acceptable alternatives under federal regulations appear ill-suited for that segment of the industry with the greatest need. The self-bonding alternative embodies qualification criteria that are too rich for the smaller operators. The most readily available alternative remains the use of some form of collateral, usually cash or its equivalent. The absence of readily available alternatives other than cash equivalents forces the small operator to tie up considerable capital that otherwise could be used for ongoing mining and reclamation activities, and this further exacerbates the company's difficulties in accessing surety coverage for future operations.

The unintended consequence from the lack of available bonding alternatives could be the creation of a built-in default base among certain segments of the industry. To avoid this result, policy makers should consider expanding the alternative bonding mechanisms accessible to those with acute need. States have recently demonstrated greater interest in various risk-sharing mechanisms such as bonding pools and sinking funds. These arrangements could hold the greatest potential for a near-term solution to the bonding crunch.

The Small Operators Assistance Program (SOAP) under SMCRA is another statutory provision that demands the attention of policy makers. In its present form, SOAP offers little more than a Band-Aid to stem a massive hemorrhage. While well intended and perhaps useful in its original mandate, two reasons largely explain the dramatic downturn in the number of applicants for assistance under SOAP. First, the number of operators who potentially qualify, those with annual production of one hundred thousand tons or less, has declined sharply since SMCRA was enacted. Second, the type of assistance offered, primarily hydrologic data analysis, remains too narrow to provide meaningful relief from the costs associated with regulatory requirements. Assistance, such as bonding, directed toward SMCRA requirements that have emerged as barriers to market entry, may prove more meaningful to the small operator.

Legislative action is required to correct both shortcomings. While recent amendments to SMCRA increased the eligibility criteria to three hundred thousand tons of annual production, they did not address the second shortcoming of the program. Congress will certainly look to policymakers in the state and federal agencies, as well as in the industry, for guidance. Again, an understanding of changes in the industry and the underlying reasons for these changes will enable policymakers to target their suggestions to meaningful revisions of this aspect of SMCRA.

Often overlooked in the dynamics of SMCRA policy development are the needs and preferences of various localities and individual landowners. No doubt, SMCRA provides ample opportunity for individuals to participate in the formal decision process. Yet, even under the best of circumstances, policy choices made at the national level for the general public good inevitably preclude other choices traditionally left to private accommodation among the directly affected parties.

A persistent criticism of SMCRA is that, in some respects, the statute usurps the prerogatives of the landowner. To some, the ubiquitous requirement to return the land to its approximate original contour precludes vast opportunities to create more economically productive areas, particularly in Appalachia. Another emerging concern among landowners is the extended period before final bond release. For landowners, this requirement unduly hampers the return of the land to economic use well beyond the time the land has reached productive capacity. Landowner sentiment led the North Dakota legislature to enact a resolution urging amendment of the federal law to allow each state to determine the appropriate time frame for bond release.

Shifts in environmental values will likewise necessitate changes in SMCRA policy. For instance, concern over depletion of the nation's wetlands raises questions about policies related to permissible postmining land use choices. It has been well documented that agricultural activities account for most of the loss of inland wetlands. Through its reclamation capabilities, the coal industry has a unique opportunity to create wetlands and introduce bio-diversity into areas that have become biological monocultures. However, SMCRA policy actually impedes industry potential to augment the nation's wetlands inventory. Present policy leaves little choice but to return croplands to the same postmining land use, and it appears at odds with today's national environmental priorities. It seems ironic that SMCRA policy precludes coal producers from converting cropland to wetlands and wildlife habitat, while the government expends substantial sums in an effort to discourage others from using these same lands for agriculture. In regard to the future efficacy of policy development under SMCRA, changes in local needs and national environmental values will certainly become as important as changes in the regulated industry.

Primacy

The federalist scheme that Congress chose to implement SMCRA is well suited to maintain a regulatory program responsive to an evolving coal industry. By leaving the regulatory function to the states, Congress took the view that they were most capable of developing and applying regulatory policy relevant to the peculiar needs and characteristics of local industry. One cannot expect a singular regulatory approach from the federal government to be responsive to, and thus effective in, regulating a diverse industry that operates in a diverse environment. Instead, the federal government should implement a national policy that affords the states sufficient flexibility to devise specific applications relevant to their local conditions and industry structure. Anything less not only abridges the deliberate allocation of authority, but also compromises the integrity of the implementation scheme by increasing the prospects for conflict because of uncertainty.

The federalist scheme has been severely tested during the first decade under SMCRA. An accurate assessment of primacy as an effective implementation scheme has been hampered by several factors. First, in view of the turbulence associated with the model federal regulations, no readily available baseline exists to measure performance. Second, both the regulators and the regulated encountered difficulties in the transition from the initial phase of shared federal and state responsibilities to the permanent phase when the states assume exclusive regulatory jurisdiction over mining activities. Finally, the transition to primacy coincided with a period of significant change in industry structure.

All three factors introduced substantial uncertainty into the program implementation. However, the reluctance of the federal agency to hand over completely the regulatory reins to the states compounded the uncertainty that the states experienced as a result of the other factors not exclusively within their control. The de facto presence of two regulatory masters in the field places coal operators in the middle of state-federal feuds over application of the law. Even the soundest compliance strategy cannot withstand the vagaries of an unstable regulatory climate. Moreover, residual uncertainty severely compromises the statutory equation that places a premium on advance planning and sound resource management.

The absolute elimination of conflict appears remote inasmuch as sharing responsibilities under the same program inevitably leads to duplication and tension. A more meaningful endeavor might be to establish a framework that would minimize the likelihood of day-to-day conflict between the parties entrusted with implementing SMCRA, and this, in turn, would afford the regulated community greater certainty under the program.

Despite what could appear as evolution toward a more consolidated industry structure, the coal industry remains anything but monolithic in character. A diverse industry profile still exists, not only among the different coal regions, but also within the regions themselves. Of course, the principal factors that Congress cited as reasons for placing regulatory responsibility with the states—distinct terrain, climate, and geologic conditions—remain unchanged. All things considered, the state primacy scheme of SMCRA remains a superior mechanism to deliver responsive and relevant regulatory policy for a diverse industry. It is better than a formula that reposes all responsibility with the federal government.

Regulatory policy under SMCRA has influenced the coal industry's development. However, the structural changes observed in the industry largely resulted in response to turbulent market forces. The ability of policy makers to observe and understand changes in the industry will greatly enhance the effectiveness of future policy development under SMCRA.

REFERENCES

Energy Information Administration. 1988. *The Changing Structure of the U.S. Coal Industry 1976–1986*. DOE/EIA - 0513, Washington, D.C.: U.S. Government Printing Office.

———. 1989. *Coal Data: A Reference*. DOE/EIA-0064(87). Washington, D.C.: U.S. Government Printing Office.

———. 1987. *Coal Data: A Reference*. DOE/EIA-0064(85). Washington, D.C.: U.S. Government Printing Office.

———. 1980. *Coal Data: A Reference*. DOE/EIA-0064(80). Washington, D.C.: U.S. Government Printing Office.

———. 1990. *Coal Distribution, January-December 1988*. DOE/EIA-0215(89/4Q). Washington, D.C.: U.S. Government Printing Office.

———. 1989. *Coal Production 1988*. DOE/EIA-0118(88). Washington, D.C.: U.S. Government Printing Office.

———. 1986. *Coal Production 1985*. DOE/EIA-00118(85). Washington, D.C.: U.S. Government Printing Office.

National Coal Association. 1990. *Coal Data: 1990 Edition*. Washington, D.C.: National Coal Association.

———. 1988. *Coal Industry Employment/Production*. Washington, D.C.: National Coal Association.

Quinn, Harold. 1989. "Coal Resource Development and Land Use Planning: The Demands of SMCRA." *Natural Resources and Environment* 3(4): 24.

U.S. Department of the Interior. 1990. *Federal Coal Management Report*. Washington, D.C.: U.S. Government Printing Office.

5 The Surface Mining Control and Reclamation Act and Environmental Groups

James M. McElfish, Jr.

Environmental organizations have played major roles in how the Surface Mining Control and Reclamation Act (SMCRA) has been implemented. These groups have used not only the array of citizen-oriented tools provided in the law itself, but also their political options to pursue their objectives. Despite signal successes by a number of environmental organizations in redirecting the focus of federal and state funds and staff resources, experience has shown that even the most sophisticated environmental activism can only partially affect on-the-ground results. The influence of these environmental groups upon state and federal efforts has been effective as a partial guarantor of SMCRA implementation.

A CITIZENS' LAW

Probably more than any other United States environmental law, SMCRA was the product of grassroots citizen activists. The law did not originate with Washington-based professional environmental organizations, nor did it arise as the logical outgrowth of typical broad-based middle-class environmental concerns like those that led to legislation on clean air, clean water, or even hazardous waste dumping.

Rather, SMCRA came about because a coalition of diverse coalfield citizen organizations recognized the infeasibility of dealing with coal mining—in particular, strip mining—on a state-by-state basis. These citizen organizations had come to perceive that the coal industry had a virtually unbeatable hold on some state legislatures in the 1960s. In the 1970s, an array of grassroots organizations from Appalachia and the West joined forces with church organizations, organized labor, and the professional environmental community in Washington, D.C., to work for the enactment of a federal strip mining law (Vietor 1980).

The critical leadership and assistance provided by the national professional environmental organizations cannot be overlooked; they provided the staff and publications resources that helped make the law possible. But the reason the law came to exist at all was because of the grassroots activists.

The effort to enact a federal strip mining law took nearly a decade. The political struggles, which included efforts to ban strip mining, as well as strenuous industry resistance to any form of federal regulation, were intense. Both sides used sophisticated lobbying techniques and waged massive public relations campaigns. The effort survived two presidential vetoes of mining regulation bills in 1974 and 1975, before President Jimmy Carter signed SMCRA on August 3, 1977, at a Rose Garden ceremony attended by the leaders of the environmental community.

Much to the disappointment of many citizen advocates, SMCRA, as finally enacted, provided that primary responsibility for its implementation and enforcement would lie with the states. This was particularly disappointing to the environmental organizations because much of the impetus for the law was the perception that the states could not be trusted to regulate. Such a perception, however, did not prevent Congress from following the by then familiar pattern of cooperative federalism—state primacy subject to oversight by a federal agency. In this case, the federal agency was the U.S. Department of the Interior. Congress's choice of oversight agencies was not entirely to the environmentalists' liking either. The Interior Department was perceived by them as beholden to mineral development interests and not primarily interested in environmental protection (Vietor 1980, 91).

Given the grassroots origins of the bill, coupled with its proponents' distrust of the regulated industry, the states, and the federal agency charged with oversight, there is little wonder that the environmental groups worked hard to build into the bill numerous opportunities for citizen involvement. These included citizen complaints, citizen-accompanied inspections, citizen petitions for unsuitability designations, citizen suits against state and federal regulators, citizen suits against coal operators, citizen participation in permitting, citizen review of bond releases, damage suits, attorney fees, and other provisions. Congress noted that "an increased opportunity for citizens to participate in the enforcement program [is] necessary to assure that the old patterns of minimal enforcement are not repeated" (U.S. Senate 1977).

The political struggles between the coal industry, which had bitterly resisted any form of federal regulation, and citizens groups, which had pressed for the law, continued well beyond SMCRA's enactment. While citizens groups found things to dislike about SMCRA, the coal industry disliked it even more. The industry, supported by a number of states that opposed federal regulation, immediately launched constitutional and other legal challenges to the law.

While the Supreme Court unanimously upheld the law in 1981 (Hodel v. Virginia Surface Mining and Reclamation Ass'n, 452 U.S. 264 [1981]), conflict over implementation continued over the ensuing decade. One measure of the continuing split between the industry and citizens is the fact that many of the regulations that implemented the program are still in litigation today. In 1977, the citizen-industry struggle simply shifted from Congress to the courts, to federal and state regulatory agencies, and to congressional oversight committees (Shover, Clelland, and Lynxwiler 1986).

With the advent of the Reagan administration in 1981 the conflict intensified. That administration—with its deep distaste for government regulation, its preference for states' rights, and its prodevelopment appointees at the Department of the Interior and the Office of Surface Mining—made no pretense about which side of the conflict it was on. James Watt, President Reagan's first secretary of the interior, had been director of the Mountain States Legal Foundation, where his last act was to file an *amicus* brief with the Supreme Court contending that SMCRA violated the U.S. Constitution. James Harris, the new director of the Office of Surface Mining (OSM), had led the fight in Indiana against SMCRA. J. Steven Griles, the new deputy director of OSM (and later assistant secretary for land and minerals) came from the Virginia regulatory program, which had also vigorously opposed the federal legislation.

The surprising thing about citizen participation in this postenactment struggle is not that it occurred. Clearly, the citizen proponents of SMCRA expected from the beginning to have to hold both government and industry to account. Rather, the surprise is in the forms that citizen action took. Citizen involvement in SMCRA implementation has only rarely been mine-by-mine or based on the specific citizen-participation tools in the law. Instead, citizens groups have launched sophisticated multipronged efforts to achieve structural and programmatic changes in governmental implementation. These efforts, as outlined below, have been remarkably effective in shaping government allocations of resources, even over strenuous agency resistance. Their on-the-ground effects are less clear, but still significant.

THE ORGANIZATIONS

The effort to enact SMCRA involved more than 30 national and regional environmental and citizens groups. These ranged from rural citizens' committees, to religious organizations like the Commission on Religion in Appalachia (22 denominations) and the National Catholic Rural Life Conference, to regional federations like the Northern Plains Resources Council, to national professional environmental organizations like the Sierra Club, the National Wildlife Federation, the National Audubon Society, the Environmental Policy Center, and the Izaak Walton League. Many of the organizations banded together to work for the bill through the Coalition Against Strip Mining.

After SMCRA was enacted, however, the organizational constituency began to change. Many of the national organizations withdrew to their traditional areas of interest, and the task of assuring implementation fell primarily to the grassroots organizations. The notable exceptions to this trend were the National Wildlife Federation and the Environmental Policy Center (now Friends of the Earth), which maintained a relatively small, but critical, presence in coal mine regulation into the 1990s.[1] With the important exception of these two national organizations, however, local citizen groups have carried on most of the effort to assure implementation of the law.

The reasons most of the professional environmental groups withdrew from the SMCRA arena are complex. In part, the pullback may have been the natural reaction of organizations to the conclusion of a legislative struggle that had lasted years longer than many of them had anticipated. Also, the result had been favorable—with the enactment of the law, "the good guys had won." The fact that the Carter administration appointed environmentalists to positions at the departments of Interior and Justice helped reinforce that impression. Finally, the urban middle-class membership of many of the national organizations and the increasingly specialized focus of their professional staffs militated against their continued allocation of significant resources to an issue that (at least in implementation) was relatively arcane and affected primarily rural coalfield residents. The striking thing is not that so many national organizations abandoned the issue after 1977, but that two of them chose to remain.[2]

Even the onset of the Reagan administration did not result in the other national organizations refocusing their efforts on SMCRA. While these organizations devoted plenty of attention to Interior Secretary James Watt, and while a number of them sponsored membership campaigns with him as their *bête noire,* issues such as offshore oil leasing, antiwilderness activities, and the hazardous waste scandals emerging at the Environmental Protection Agency attracted much of their attention.

At the grassroots level, a variety of organizations pursued SMCRA activities. Those that have remained active since the 1970s include the Resource Councils of the western states (e.g., Northern Plains Resources Council), the Illinois South Project (renamed Illinois Stewardship Alliance in 1990), Save Our Cumberland Mountains (Tennessee), the West Virginia Highlands Conservancy, Kentuckians for the Commonwealth (formerly Kentucky Fair Tax Coalition), and the Kentucky Resources Council. Some groups active in the 1970s disappeared in the 1980s, while others emerged. In the 1980s, several organizations like the Virginia Citizens for Better Reclamation (VCBR), and the 76-year-old Council of the Southern Mountains folded their operations. Like most grassroots groups, those pursuing coal mining issues have waxed and waned, depending upon the urgency of local problems and the presence or absence of a critical mass of dedicated volunteer members (Hays 1987, 62).

On the other hand, new coalfield organizations appeared throughout the 1980s. Many of the newer organizations are interested primarily in the adverse effects of subsidence and the destruction of domestic water supplies by underground mining. These groups include the Dickenson County Citizens Committee (Virginia), Protect Our Water and Environmental Resources (Pennsylvania), Citizens Organized Against Longwalling (Ohio), and Citizens Against Water Loss due to Mining (northern Appalachian states).

The grassroots organizations, new and old, organized the Citizens Coal Council in the late 1980s to serve as a forum to pursue SMCRA issues. In 1990, the Council opened a Washington, D.C., office with one staff member.

In any discussion of the effect of environmental groups on SMCRA, it is important to bear in mind that two types of organizations exist: the national professional environmental organizations and the grassroots groups. Both types have played critical roles in SMCRA implementation at the national and local levels. The national organizations have become involved in local issues, and the grassroots groups have sought to influence national policy. This combination of focus has occurred in part because the two national organizations active in the field—the National Wildlife Federation and the Environmental Policy Center (now Friends of the Earth)—have deliberately devoted a substantial portion of their efforts to supporting the efforts of the grassroots organizations. The national organizations have not pursued an agenda divergent from that of the coalfield groups. Thus, the discontinuity between Washington-based groups and grassroots groups that is frequently observed in other fields of U.S. environmental endeavor—hazardous waste, for example—has not emerged in the SMCRA field. The reasons for this difference in the SMCRA area are both institutional and personal. The two Washington-based organizations have sought out and hired citizen-activist staff for their coal programs, and the grassroots groups have frequently recognized the need for (and called upon) Washington-based support in their struggles with state regulatory agencies.

THE ENVIRONMENTAL AGENDA: DEFENDING THE 1977 GAINS

A crucial effort of the national environmental organizations in the 1980s was to prevent weakening of the 1979 SMCRA regulations. In 1981, James Watt took office as secretary of the interior and pledged to rewrite the entire set of SMCRA regulations. This task was completed by 1983. The National Wildlife Federation took the lead in challenging the rewritten regulations and achieved substantial success in front of U.S. District Court Judge Thomas Flannery. Judge Flannery has spent much of his career on the federal bench presiding over the aptly, if unintentionally ironically, named In re Permanent Surface Mining Regulation Litigation. Notable wins by the National Wildlife Federation included limiting variances from the approximate original contour requirements to steep slope areas, redefining previously mined areas, requiring repair of subsidence-damaged structures (reversed on appeal in 1991), and strengthening criteria that govern mining in national forests.

Environmental organizations were much less effective in their attempts to challenge state program approvals. In 1981, Secretary Watt relaxed the regulatory standard for approval of state program regulations. He replaced a rule that required state regulations to be "no less stringent than" the federal program regulations with one that required them to be "no less effective than" the federal regulations (46 *Federal Register* 53,384 [Oct. 28, 1981]). The new standard, coupled with OSM's policy (initiated by the Carter administration) of approving state programs "conditionally" where they did not meet the federal standards, was viewed by the environmental organizations as a threat to the stringency and integrity of state programs.

The Environmental Policy Institute (EPI), the 1980s sobriquet of the Environmental Policy Center, recognized the critical importance of state program approvals to SMCRA implementation. Because it had no direct way to influence the Reagan appointees at the Department of the Interior with respect to such approvals, EPI focused its efforts on litigation. By law, challenges to state program approvals could be brought only in the federal district courts in the state capitals of the respective states. This litigation proved to be a formidable task for a national organization. The group needed to seek out local plaintiffs to bring the challenges and local lawyers to handle them. It needed to craft arguments that would appeal to judges who had little familiarity with the surface mining program. Although a number of the suits were partially successful, for the most part the courts upheld the programs, deferring to the presumed expertise of the secretary of the interior and the state legislatures and agencies.

The Environmental Policy Institute's ability to challenge state programs effectively was further hindered by a strategem adopted by a majority of the eastern coal states. Many of these states successfully delayed submission of their programs to the Department of the Interior for approval in order to continue under the more relaxed "interim regulatory program." Section 503(d) of SMCRA granted the states a one-year deferral from submission of state programs if they were enjoined from doing so by "a court of competent jurisdiction." Beginning with Indiana, most of the eastern states were so enjoined in suits filed in state courts by coal operators' associations. In most states, the request for the injunction was not opposed by the state defendants. The delay occasioned by these 503(d) suits further taxed the EPI's ability to maintain a sustained effort.

The conditional approvals of state programs also remained a problem substantially beyond the ability of the environmental groups to address. In some states, conditions imposed on initial program approval were not satisfied until the late 1980s or even the early 1990s. Thus, the programs in these states failed in certain respects to meet the federal baseline. The environmentalists had no obvious avenue to speed the process of upgrading these programs. Litigation would have been difficult, particularly as the baseline federal regulations were themselves constantly changing due to regulatory revisions and court decisions.

In sum, the environmental organizations were highly successful in blocking or overturning the regulatory rewrites undertaken by Secretary Watt. They were far less successful in affecting the content of state programs.

THE ENVIRONMENTAL AGENDA: "THE BEST DEFENSE IS . . . "

While the defense of the regulations from Secretary Watt's rewrite and the attempt to hold the line on state program approvals could have occupied environmental organizations to the exclusion of all else, it did not. The National Wildlife Federation bore the brunt of the regulations effort. In the meantime, the environmental groups went on the offensive.

The idea of an active, rather than reactive, environmentalist approach to SMCRA actually predated the Watt years. In 1979, Tom Galloway, at the Washington, D.C., Center for Law and Social Policy, sued the new Office of Surface Mining on behalf of the Council of the Southern Mountains (Council of the Southern Mountains v. Andrus, Civ. Action No. 79–1521 [D.D.C.]). The suit charged OSM with failing to conduct the number of inspections required under the interim program (at least two per year at each mine site) and with other enforcement deficiencies. OSM settled the suit and increased its inspection and enforcement efforts, using the settlement to extract greater federal funding and to obtain greater institutional support within the Department of the Interior for SMCRA implementation. This suit became a paradigm for environmental organizations' approaches to the SMCRA program in the succeeding decade. The suit sought structural and programmatic changes, and it inserted the environmental organizations firmly into the agency's decision-making process. OSM could not treat the holder of an injunction as just another outside interest group. An organization with an injunction has real power to obtain access to government records, to second-guess government decisions, and to influence policy.

The approach used in 1979 was used repeatedly throughout the 1980s. The environmental organizations were even more successful with their structural assaults on perceived defects in implementation in the 1980s when OSM was controlled by openly hostile political interests. The "two-acre" issue, the "mega-bucks" case, and the Kentucky litigation illustrate this approach.

The Two-Acre Issue

In 1977, Congress provided certain exemptions in SMCRA for coal extraction deemed environmentally inconsequential. One such exemption, added over the objection of environmental advocates, allowed operators who affected two acres or less to avoid regulation under the law. Congress described the provision as one designed to accommodate "mom and pop" operations that caused minimal effects.

As early as 1979, some operators devised methods of conducting large, unregulated mining operations under the guise of the two-acre exemption. These methods included having separate corporations mine adjacent two-acre sites with the same equipment, setting up a string of "two-acre" operations separated by gaps of unmined land along a single coal seam, and deeding haul roads to cooperative county governments to keep acreage below two acres. OSM promulgated a rule in January 1981 to deal with abuse of the exemption, but the rule was suspended by Secretary Watt. The coalfield environmental groups pursued a multipronged strategy to attack the problem. First, they documented the problem through permit files and site visits. Then, Save Our Cumberland Mountains, the Council of Southern Mountains, and Virginia Citizens for Better Reclamation filed suit in Washington, D.C., alleging a nationwide breakdown in "two-acre" enforcement. They initiated Congressional inquiries, and they pressed for OSM enforcement. They also embarked on a publicity campaign to highlight the abuses. In addition to popular media, including network television, their attorneys published articles in scholarly journals that documented the abuses (Galloway and FitzGerald 1985).

Although the organizations lost the case on procedural grounds in the district court and in the court of appeals, they persuaded the court of appeals to grant a rehearing before the full court (Save Our Cumberland Mountains v. Clark, 725 F. 2d 1422 [D.C. Cir., 1984], vacated *en banc* [1984]). Before the rehearing occurred in 1984, the government settled, pressured by a combination of publicity, congressional investigations, and other prongs of the environmentalists' approach. While the litigation was pending, OSM took over two-acre enforcement from the state programs in Virginia and, later, Kentucky. After the settlement, OSM took additional measures.

Finally, in 1987, Congress repealed the two-acre exemption, primarily because of the record developed by the environmental organizations. By the time the issue reached Congress, the call for repeal was supported formally, not only by environmental groups but by the states, industry, and OSM. The bill passed easily.

"Mega-bucks"

During the early years of the interim program, OSM was unable to keep up with assessment and collection of penalties against operators who ignored both notices of violation and cessation orders. These penalties mounted at a statutory minimum rate of $750 per day, until some of them reached hundreds of thousands of dollars. These amounts, popularly referred to as "mega-bucks," often could not be collected from small operators. In order to deal with this problem, OSM promulgated a regulation that cut off the daily assessment at 30 days, or $22,500 per violation and that committed OSM and states with approved programs to take "alternative enforcement action" within 30 days of the cutoff. The alternative actions could include permit revocation, bond forfeiture, civil penalties against individuals, or criminal enforcement (30 CFR 723.15, 845.15). By the end of the Carter administration, however, OSM had neither collected the outstanding civil penalties from the scofflaw operators nor implemented alternative enforcement actions.

In 1981, when it became clear that Secretary Watt did not intend to pursue these operations or take alternative enforcement action against the violators, Save Our Cumberland Mountains and the Council of Southern Mountains filed suit. In late 1982 and early 1983 they received favorable rulings from U.S. District Judge Barrington Parker, who enjoined OSM to collect the outstanding penalties and to take alternative enforcement actions against approximately eleven hundred operations (Save Our Cumberland Mountains v. Watt, 550 F. Supp. 979 [D.D.C. 1982], reversed 725 F. 2d 1434 [D.C. Cir. 1984], reversal vacated *en banc* [1984]). At the same time, Save Our Cumberland Mountains and the Council of Southern Mountains, together with the national organizations, lobbied Congress to increase funding for OSM to carry out the terms of the order. In 1983, Congress appropriated an additional $1.1 million to OSM and directed the agency to carry out the Parker order. Additional sums were appropriated in later years.

Although in 1984 a panel of the court of appeals initially reversed Judge Parker's injunction on procedural grounds, the full court of appeals vacated the reversal order. The Department of the Interior settled with the environmental organizations in 1985. The settlement was far more comprehensive than the original order had been. It provided not only for OSM to evaluate the past cases and attempt to collect penalties and impose alternative enforcement, but also for OSM to construct a sophisticated computer system (the Applicant/Violator System or AVS) to ensure that past violators (or their owners and controllers) could not obtain new mining permits. This settlement transformed the case from one about past practices into a structural approach to prevent future abuses. The settlement came about in part because of the environmental organizations' increasing use of congressional pressure in conjunction with the injunction.

Congress appropriated substantial sums for the AVS over the succeeding years, and OSM devoted significant staff resources to the project. Unfortunately, despite the striking success of the citizens groups and the eventual expenditure of some $15 million, the system had serious technical flaws. Once again, the organizations turned to Congress. In 1987, Representative Morris Udall, chairman of the House Interior Committee, requested the

General Accounting Office (GAO) to investigate the operation of the AVS. Meanwhile, the National Wildlife Federation had developed its own AVS on a microcomputer, to assist coalfield citizens groups in evaluating mining applications. The GAO concluded in 1989 that the National Wildlife Federation's skeletal AVS system outperformed the mainframe OSM system (GAO 1989). The citizen organizations filed a motion to have OSM held in contempt of court, and in 1990 a settlement was reached to upgrade the system. The factors driving the settlement were not simply the obvious weaknesses in the federal AVS, but also the need by the Bush administration (and its new appointees) to restore credibility with the House Interior Committee.

Although the AVS had defects, its establishment changed the terms for enforcement in the late 1980s. OSM and the states both began to focus on the SMCRA "permit block" that prevents violators and their affiliates from obtaining mining permits.

Although the remedy prescribed in the settlement of the mega-bucks case was far different from what one might have predicted by reading the complaint filed in court in 1981, the environmental groups achieved what they were after—structural reform. Indeed, the purpose in filing the case never had been the collection of overdue civil penalties; rather, that was the vehicle for reform of OSM's enforcement focus. Alternative enforcement and, later, the permit block, were the actual objectives. These powerful tools in SMCRA had been largely unused by OSM and most state regulatory authorities. The mega-bucks case did a great deal to prevent SMCRA enforcement from lapsing into a traditional version of EPA-style enforcement (issue the enforcement document, collect the civil penalty). The citizen groups forced the federal agency to focus on the permit, the reclamation bond, and the personal remedies (individual civil penalties and criminal enforcement).

The mega-bucks issue also demonstrated the power of the multipronged strategy of the citizen groups. The combination of congressional oversight, congressional funding, General Accounting Office investigations (GAO 1987), and litigation produced significant changes in implementation and enforcement.

Kentucky

As though the environmental groups did not have enough irons in the fire, in 1986 the National Wildlife Federation and other organizations filed suit against the Commonwealth of Kentucky, citing the state's failure to carry out significant parts of its approved program (National Wildlife

Federation v. Miller, Civ. No. 86–99 [E.D. Ky.]). Once again, the citizens did not simply focus on particular breakdowns, they attempted to identify and remedy structural problems such as inadequate staffing and training and insufficient technical resources. The groups simultaneously filed suit against OSM for failing to oversee Kentucky's program. While the latter suit was dismissed on procedural grounds, the first suit generated substantial results. First, OSM and the Kentucky Natural Resources and Environmental Protection Cabinet immediately signed an agreement to upgrade Kentucky's program. This immediate effect of the litigation resulted from the governmental agencies' desires to put themselves in the best posture to have the citizen litigation dismissed. The litigation against Kentucky was not dismissed, however, and the state entered into settlement discussions with the environmental organizations.

As the organizations pled their case in the federal court in Frankfort, they also pled it to the House Interior Committee in Washington, D.C. After many months of intensive negotiation, the environmentalists succeeded in convincing Congress that Kentucky would settle the case if it could get federal money to upgrade its program and convincing Kentucky that Congress would appropriate money if the state would just settle the case. Both were persuaded, and in September 1987 the case was settled. Congress appropriated nearly $12 million over three years to implement the settlement.

The Kentucky settlement surpassed any other citizen-suit outcome in U.S. environmental law in terms of its specificity and the level of control it granted environmental organizations over governmental functions. The settlement agreement established the specifics of staffing the agency, purchasing equipment, training personnel, and setting enforcement priorities. It allowed citizens to focus state resources on particular sites and to control enforcement discretion. For example, among other provisions, it spelled out (in full-time equivalents) precisely how many persons with what types of skills would be assigned to particular functions; it provided for helicopter overflights of mine sites at a specified number of hours per year; and it established computer tracking systems to share data with the plaintiff organizations. The settlement, in effect, made the plaintiff organizations partial comanagers of the Kentucky program, in exchange for the additional federal dollars obtained through the organizations' lobbying efforts.

The striking success of the Kentucky litigation made a significant difference in the Commonwealth's enforcement of SMCRA. Indeed, an editorial in the leading industry trade publication acknowledged that the program had been improved by the settlement and called on regulators and industry to put their "regulatory houses in order" to forestall future suits in other states (Sanda 1990). However, the major difference made by the Kentucky suit was the devotion of additional resources to the task. These resources most likely could not have been obtained by OSM or Kentucky seeking a special Congressional appropriation. What OSM and Kentucky could not have done was done by citizen litigation coupled with a sophisticated understanding of Capitol Hill. Struggles remain over Kentucky enforcement. As the federal money runs out, the environmental organizations may find that they have less leverage. Nevertheless, the Kentucky litigation marks a highly sophisticated partial solution to problems of breakdown of SMCRA implementation in the states.

In 1988, the National Wildlife Federation and other organizations filed a similar suit against West Virginia (Save Our Mountains v. Faerber, Civ. No. 2:88–1609 [W.D.W.Va.]). West Virginia entered into a partial settlement of the case. The limited availability of additional federal funding and West Virginia's own fiscal woes may, however, prevent a full repetition of the environmental organizations' Kentucky success.

The Multipronged Approach

These three cases illustrate the sophistication of environmental groups in their actions to produce structural reforms in SMCRA implementation. In these efforts, the national groups worked closely with the grassroots groups. Sometimes one would take the lead, sometimes another. With respect to the mega-bucks litigation, the National Wildlife Federation, while not a named plaintiff, devoted considerable resources to establishing its own shadow applicant/violator system. The National Wildlife Federation's system was used to demonstrate the inadequacies in the OSM system and was made accessible to the grassroots groups so they could perform their own checks on mining permit applicants. This type of cooperation and division of labor has greatly enhanced the collective effectiveness of the groups interested in SMCRA.

One of the chief architects of the comprehensive structural approach used by the citizen groups has been Tom Galloway, a Washington, D.C., attorney. Galloway has represented virtually all of the groups at one time or another, and his planning has frequently enabled the organizations to apply the most effective tools to achieve their objectives. He has been successful, and ubiquitous, enough to draw the wrath of much of the coal industry. He is mentioned here, not to downplay the importance of the organizations, but to highlight the importance of a strategist in wielding relatively limited organizational resources to maximum effect.

One particularly effective approach has been to ask members of Congress to request GAO investigations of various problems. Invariably, with help from the environmental organizations, GAO has documented the abuses identified. In the 1980s, GAO investigated OSM on more than a dozen occasions. Likewise, the House Interior Committee and the Government Operations Committee have investigated OSM numerous times (see U.S. House Committee on Interior and Insular Affairs 1985; and U.S. House Committee on Government Operations 1987).

The success of the environmental organizations in all of these endeavors depended more than anything else on the support of Congressman Morris Udall. Building on the relationships forged during the struggle to enact a surface mining law, the organizations in Washington and in the coalfields were able to count upon him to make inquiries into OSM's performance, to seek GAO scrutiny of field conditions, and to add monies not requested by OSM to its annual appropriation. (This last factor is one of the least noticed and most important facts of 1980s implementation.) Congressman Udall's personal support for the law, coupled with the respect he garnered among his colleagues in Congress, was the single greatest asset the environmental organizations had.

It is evident that the organizations' multipronged strategy throughout the 1980s depended heavily on congressional help to support the litigating victories (and to stem the litigating setbacks). It was particularly important during the Reagan administration, when the groups' influence on the bureaucracy was at an all-time low.

It remains to be seen what the new dynamic will be. There has been a more mainstream and open OSM under the Bush administration, while Congressman Udall assumed a more limited role in SMCRA oversight. Before retiring in 1991, he relinquished the subcommittee with jurisdiction over SMCRA to Congressman Rahall of West Virginia. Nevertheless, the environmentalists' multipronged approach appears to offer sufficient flexibility for the groups to continue to influence SMCRA implementation.

COALFIELD LEGISLATIVE EFFORTS

The coalfield citizen organizations did not confine themselves entirely to SMCRA issues or to Washington-based initiatives. Two efforts in the 1980s illustrate the impact of the grassroots groups at the state legislative level.

In 1984, groups led by the Kentuckians for the Commonwealth persuaded the Kentucky legislature to enact the Mineral Deed Act. This act was designed to impose a rule of construction on broad form deeds. The act established a rebuttable presumption that mineral deeds that were silent on the mining method to be used authorized only those methods commonly known to be in use in that area of Kentucky at the time of the conveyance. In short, the act was an attempt to assure that deeds executed in the early 1900s and before could not be held to authorize surface mining, absent express provisions in the deed. In 1987, however, the Kentucky Supreme Court struck down the law as a violation of the Kentucky Constitution (Akers v. Baldwin, 736 S.W. 2d 294 [Ky. 1987]).

Undeterred, the citizens organizations launched an effort to amend the Kentucky Constitution. In 1988, the voters enacted the Broad Form Deed Amendment by a vote of 882,960 to 187,119. The amendment substantially realigned the bargaining power between mineral and surface owners with respect to older mineral conveyances. Its enactment represented a triumph for the citizens groups. A challenge to the constitutionality of the amendment under the U.S. Constitution has followed, so it remains uncertain whether the citizens will ultimately prevail (Bratt and Greenwell 1989–90). Nonetheless, as demonstrations of legislative determination and public clout, the Kentucky legislation and the constitutional amendment reflect significant shifts in public opinion, brought about by the efforts of coalfield citizens organizations.

Save Our Cumberland Mountains also pursued legislation in the 1980s. In 1987, it persuaded the Tennessee legislature to enact a Surface Rights Law. The law allows surface owners to reunite mineral interests with the surface estate, where the mineral owner has abandoned the interest. It required all mineral owners to register their interests with county assessors by July 1, 1990. Interests not registered are subject to reunion with the surface estate upon petition by the surface owner. The law was designed to eliminate confusion over land titles and to assure that mineral interests upon which taxes have not been paid and other incidents of ownership not exercised could be restored to surface owners. Save Our Cumberland Mountains not only sponsored the law and saw to its passage, it followed up with numerous workshops to assure that coalfield surface owners could take advantage of the law's provisions.

These efforts on the state level represent sophisticated use of legal and political tools by local groups to achieve their ends. These groups have not limited themselves to responding to specific threats or to exercising the rights provided by SMCRA. Rather, they have sought structural changes.

Other state-level efforts under way in the early 1990s include those by several groups in Appalachian states to obtain legislation that protects domestic water supplies from the adverse affects of underground mining. Section 717 of SMCRA requires surface miners to replace water supplies they destroy or diminish, but it does not impose such a requirement on underground mine operators. State-level efforts to enact water replacement legislation are a significant focus for coalfield groups and, along with struggles over subsidence, will evidently comprise the critical issues for the 1990s.

MINE-BY-MINE ACTIONS

In general, environmental organizations have focused their efforts on structural and programmatic changes in the administration of SMCRA. They have focused far less on the sort of mine-by-mine citizen enforcement that might have been predicted by simply reading the citizen-oriented provisions of the act.

There are several reasons for this. First, the national environmental organizations have rarely focused on individual cases under any of the U.S. environmental programs (except with respect to specific areas of the public lands). Their expertise and resources necessarily compel them to focus on programmatic issues.

Second, the coalfield citizens groups—at least those active since the 1970s—have generally been more sophisticated than their grassroots counterparts in other environmental areas (e.g., air pollution, recycling, hazardous waste). This sophistication has been shaped by their experience. Coalfield activists began in the 1970s by concluding that mining-related problems could be solved only in Washington. This perspective, while no longer entirely true given the maturing of state programs under SMCRA and changes in public perception, has nevertheless influenced the groups' strategies. While many of the groups have been active on state and local levels as described above, their history has led them to see the importance of congressional oversight, federal funding, and federal administration in solving local problems.

Moreover, attempts under SMCRA to address issues on a mine-by-mine level have confirmed the value of a structural approach. A mine-by-mine approach was insufficient to address the widespread abuse of the two-acre exemption in Virginia and Kentucky, which involved hundreds of sites.

Similarly, mine-specific efforts have frequently required a multi-pronged approach that included Washington-based resources. One of the best examples of this is the five-year struggle of Save Our Cumberland

Mountains with a single 89-acre mining operation in Bledsoe County, Tennessee. Save Our Cumberland Mountains attempted to prevent Rith Energy Company from mining a potentially acid-producing coal seam. The organization pressed OSM to retest the operation's overburden samples for their acid potential. The retest showed much higher potential for acid production than had the initial permit application under which mining had commenced. What ensued was a multiyear process in which the citizen organization used technical experts, the direct involvement of United States senators, the services of volunteer attorneys, intervention by Environmental Protection Agency officials from Atlanta, and other resources in order to persuade OSM to disapprove the mining plan and order reclamation. In the course of the struggle, the operator filed a defamation suit against the organization, its members, and its staff. This suit was dismissed by a Tennessee circuit judge one and one-half years after it was filed. While Save Our Cumberland Mountains finally prevailed and OSM ordered the necessary relief, the level of resources devoted by the organization to just this one operation gives some indication of why the mine-specific approach has not been the most common approach used by coalfield organizations.

In general, coalfield groups have pursued issues at particular mines only until rebuffed by the state or by OSM. Only rarely have citizen organizations filed suit against operators or even appealed OSM actions.

The newer coalfield groups organized in the late 1980s have not always followed this pattern. Many of them have focused on mine-specific problems rather than on strategic objectives and programmatic approaches. However, their collective involvement with the older organizations, through the umbrella Citizens Coal Council, has led them in some instances to seek structural solutions as well. This trend may continue.

CONCLUSION

Citizen organizations and national environmental groups have played significant roles in influencing SMCRA's implementation. In general these groups have preferred to seek coordinated structural and programmatic solutions and have devoted their efforts to that end. Shut out of a direct role in OSM decision making for almost a decade by the political disposition of Reagan administration appointees, the organizations employed a multipronged strategy that relied heavily on congressional resources, the General Accounting Office, and publicity to wield significant power. Mine-by-mine involvement has generally been limited in order for the organizations to focus their efforts on program-wide issues.

The shape of future involvement by these organizations remains unclear. While the multipronged approach appears certain to continue, its focus may change. The declining influence of these organizations in the House Interior Committee may be offset to some extent by greater access to OSM and to state governments.

On-the-ground effects of the involvement of environmental groups are difficult to assess. But their restructuring of state and federal implementation efforts has clearly enhanced enforcement and addressed serious problems that have environmental impacts—two-acre abuses, mega-bucks violations, weak state programs, and regulatory problems. Environmental groups have had a profound effect on SMCRA's implementation.

NOTES

1. In the 1970s, for tax policy reasons, the Environmental Policy Center created an educational arm, the Environmental Policy Institute (EPI). Later the two organizations were merged and the organization was thereafter known as EPI. In 1990, EPI, the Oceanic Society, and Friends of the Earth merged. The combined organization is now known as Friends of the Earth.

2. In fairness, it should be noted that the Sierra Club also brought several cases related to SMCRA. One early case challenged a 1981 action of Interior Secretary James Watt that relaxed the regulations for approval of state programs under SMCRA. Another series of cases, brought under the Clean Air Act, challenged EPA's failure to promulgate fugitive dust regulations for coal mines.

REFERENCES

Bratt, C., and K. Greenwell. 1989–90. "Kentucky's Broad Form Deed Amendment: Constitutional Considerations." *Kentucky Journal of Mineral Law and Policy* 5(1): 9–91.

Galloway, L. T., and T. J. FitzGerald. 1985. "Abuse of the Surface Mining Act: A Continuing Story." *West Virginia Law Review* 87(3): 627–47.

Hays, S. 1987. *Beauty, Health, and Permanence.* New York: Cambridge University Press.

Sanda, A. 1990. "Coal, Galloway, and Litigation." *Coal* 94.

Shover, N., D. Clelland, and J. Lynxwiler. 1986. *Enforcement or Negotiation.* Albany: SUNY Albany Press.

U.S. General Accounting Office. 1989. *Surface Mining: Operation of the Applicant Violator System Can Be Improved, GAO/AFMD-89–31*. Washington, D.C.: U.S. Government Printing Office.

———. 1987. *Surface Mining: State and Federal Use of Alternative Enforcement Techniques, GAO/RCED-87–160.* Washington, D.C.: U.S. Government Printing Office.

U.S. House Committee on Government Operations. 1987. *Surface Mining Law: A Promise Yet to be Fulfilled.* H.R. Rep. No. 100–183, 100th Cong., 1st Sess. Washington, D.C.: U.S. Government Printing Office.

U.S. House Committee on Interior and Insular Affairs. 1985. *Office of Surface Mining: Beyond Reclamation?* H.R. Rep. No. 99–206, 99th Cong. 1st Sess. Washington, D.C.: U.S. Government Printing Office.

U.S. Senate. "Surface Mining Control and Reclamation Act of 1977: Report together with Additional and Minority Views to Accompany S. 7." S. Rep. No. 128, 95th Cong., 1st Sess. 1977.

Vietor, R. 1980. *Environmental Politics and the Coal Coalition.* Dallas: Texas A & M University Press.

6 Federal-State Relations

Gregory E. Conrad

> The Powers delegated by the proposed constitution to the federal government are few and defined. Those which are to remain in the state government are numerous and indefinite.
>
> —James Madison

INTRODUCTION

When the Congress of the United States debated the Surface Mining Control and Reclamation Act of 1977 (SMCRA), many considered it to be one of the more complex environmental statutes ever enacted. It embodies comprehensive technical standards, extensive administrative provisions, and knotty legal procedures. Given that the act is not considered a model of legislative draftsmanship in terms of its clarity and conciseness, it is no surprise that in addition to understanding the statute's basic principles, interested parties are faced with significant matters of interpretation.[1]

Perhaps the most difficult concept embodied in SMCRA is that of state primacy, whereby the states would have the opportunity to regulate surface coal mining operations within their borders, with a minimum of federal interference. The concept is rooted in principles of comity and federalism and in the strongly held constitutional belief that the power to decide local matters is best left in the hands of local governments. And yet, the tensions that resulted in the very enactment of SMCRA tended to grow out of the perception that local governments had failed to adequately regulate the surface mining of coal and its attendant impacts on the environment.[2]

There is little doubt that the framers of SMCRA expected its participants would require a fair degree of education about the concept of primacy. What they probably did not expect, however, was the nature and extent of that learning process. The act is now in its fourteenth year of implementation—and graduation to a higher level of performance should be upon us. Yet, some would argue that the regulatory program envisioned by Congress has not passed even the elementary stages of development.

What have been the barriers to reaching the levels of achievement anticipated by Congress? Are elusive concepts to blame? Or perhaps recalcitrant participants in the process? Has too much been expected from the affected parties, given the inherent conflicts and unintended flaws of the governing legislation? Have the lessons to be learned under SMCRA been too hard, or simply hardly learned? What can be expected in the way of cooperative federalism during the next decade? All of these questions will be addressed in one way or another in the course of this chapter.

THE ROLE OF STATE GOVERNMENTS UNDER SMCRA

The Surface Mining Control and Reclamation Act established a "nationwide program to protect society and the environment from the adverse effects of surface coal mining operations" by creating a unique federal-state partnership (30 U.S.C. section 1202). The act provided for an interim transition period of direct federal regulation of surface mining by the secretary of interior, acting through the Office of Surface Mining (OSM), unless a state had its own independent regulatory program. In that case, concurrent jurisdiction would prevail. However, the act allowed each state to assume "exclusive jurisdiction over the regulation of surface coal mining and reclamation operations," generally referred to as "state primacy" (30 U.S.C. section 1253 [a]). This latter phase, the permanent program, would commence after the secretary of interior approved a state program that demonstrated the state's capability to carry out the provisions of the act. The state's program must contain rules and regulations consistent with regulations issued by the secretary of interior (30 U.S.C. section 1253[a][7]). Alternatively, if a state did not wish to control the regulation of surface mining within its borders, or if it failed to submit a satisfactory program, OSM was to develop a federal program to continue the direct regulation of mining within the state (30 U.S.C. section 1254[a]).

Once a state program is approved by OSM, that state has achieved primacy and "assume[s] exclusive jurisdiction over the regulation of surface coal mining and reclamation operations" on nonfederal lands within its borders (30 U.S.C. section 1253[a]). Currently 24 coal-producing states have received approval and are administering primacy programs. Tennessee exercised primacy for several years, but because of differences with OSM over philosophy and policy returned its program to OSM (49 *Federal Register* 38873, Federal Program for the State of Tennessee).

A review of SMCRA's history will show that the concept of primacy is rooted in principles of cooperative federalism, based on the constitutional precept that "the powers not delegated to the United States by the Constitution, nor prohibited by it to the States, are reserved to the states respectively, or to the people" (U.S. Const., Amend. X). The U.S. Court of Appeals for the D.C. Circuit recently articulated this view when it noted, "Chief among SMCRA's concerns is the need to accommodate widely varying local conditions that will not admit of a single, nationwide rule" (National Wildlife Federation v. Hodel, 839 F.2d 694, 735 [D.C. Cir. 1988]).

In fact, this policy is clearly expressed in the act itself: "Because of the diversity in terrain, climate, biologic, chemical, and other physical conditions in areas subject to mining operations, the primary governmental responsibility for developing, authorizing, issuing, and enforcing regulations for surface mining and reclamation operations subject to this chapter should rest with the States" (30 U.S.C. section 1201[f]).

In an earlier decision, the U.S. Court of Appeals for the D.C. Circuit stated that, particularly in matters involving the diverse physical conditions under which mining occurs in various regions, it was the states, with their greater familiarity with local conditions, and not the federal government, that Congress relied upon to regulate mining operations (In re Permanent Surface Mining Regulation Litigation, 653 F.2d 514 [D.C. Cir. 1981] [en banc], *cert. denied* 454 U.S. 822 [1981]).

The court also stated that "once a State program has been approved, the State agency plays the major role with its greater manpower and familiarity with local conditions. It exercises front-line supervision and the *Secretary will not intervene unless its discretion is abused*" (emphasis added) (In re Permanent Surface Mining Regulation Litigation, 653 F.2d 514, 523 [D.C. Cir. 1981] [en banc], *cert. denied* 454 U.S. 822 [1981]). The court indicated the secretary of the interior's limited role in the following passage:

> Under State programs, the State makes decisions applying the national requirements of the Act to the particular local conditions of the State. The Secretary is initially to decide whether the proposed State program is capable of carrying out the provisions of the Act *but is not directly involved in local decision making after the program has been approved.* (Emphasis added) (In re Permanent Surface

Mining Regulation Litigation, 653 F.2d 514, 518 [D.C. Cir. 1981] [en banc], *cert. denied* 454 U.S. 822 [1981])

As these and other courts have recognized, the structure of the SMCRA—with its two-stage program that transfers regulatory power from the federal government to states that formulate their own approved programs—presupposes that primacy states will assume responsibility for implementing and enforcing the act. This purposeful and deliberate sharing of powers between federal and state governments is entitled to the utmost respect. The courts and policymakers should carefully scrutinize any policy or regulation that seriously alters federal-state relations, to determine whether it is legitimate and consistent with the overall legislative intent of the act (see United States v. Del Toro, 513 F.2d 656, 662 [2d Cir. 1975]. In discerning Congressional intent, a court may consider whether an expansive interpretation of a statute would alter sensitive federal-state relationships.)

CURRENT THREATS TO PRIMACY

Primacy can have meaning only in the final analysis where deference is paid to the states to interpret their own state laws. As the court recognized in Haydo v. Amerikohl Mining Inc. (830 F.2d 494 [3d. Cir. 1987]), SMCRA itself anticipates that state courts have exclusive jurisdiction over surface mining reclamation activities within their borders. As a general proposition, then, the state program is the law within the state, and the state's interpretation of its own law should prevail. This is equally true for the adjudication of rights and remedies that arise under the act and under the state program in a primacy state. In essence, true primacy requires that states be free to interpret their own laws, especially where an understanding of local conditions is important in adjudicating a property right or in determining whether either the act or the state program has been violated.

Although Congress provided an oversight role for the secretary of the interior in terms of states implementing their approved programs, the actual components of that role were not articulated in the act. However, since the inception of state primacy, OSM has succeeded in devising an oversight role. Through aggressive enforcement OSM has effectively deprived primacy states of the full regulatory authority promised in SMCRA.

The most serious flaw in OSM's oversight policy is the agency's reflexive use of notices of violations (NOVs) as an enforcement tool, even where legitimate disputes exist between the state and OSM. These disputes arise in the context of differences in interpretation of the approved state programs, differences in subjective judgments of existing conditions at mine operations, differences of opinion as to permits issued pursuant to the approved program, and alleged deficiencies in the state program approved by the secretary of the interior. Rather than relying on the procedures provided under the act and its regulations to resolve interpretational disputes or program deficiencies, OSM has chosen to exercise its oversight authority through direct federal enforcement action. This procedure completely undermines the federalist scheme set forth in the statute and countervenes express congressional intent.

OSM's current understanding of its oversight role threatens the continued viability of primacy. Although OSM's oversight is ostensibly limited in favor of state primacy, the tension between the surface mining act and OSM's aggressive federal enforcement policy, as reflected in its regulations and policy directives, has permitted OSM to deny to the primacy states the "exclusive jurisdiction" SMCRA promises (see OSM Policy Directive No. Reg.-8 on "Oversight of State Regulatory Programs" and No. INE-35 on "Ten-Day Notices"; see also 30 CFR sections 842.11 and 843.12[a] [2]). Specifically, with regard to OSM's enforcement authority in primacy states, OSM's regulations go beyond what SMCRA allows (Clinchfield Coal Co. v. Hodel C.A. No. 85–0113–A [W.D. VA. Sept. 25, 1985], rev'd on other grounds sub nom. Clinchfield Coal Co. v. DOI, 802 F. 2d 102 [4th cir. 1986]).

SMCRA provides that OSM can order a federal inspection if the state fails to take appropriate action to correct an alleged violation within 10 days after notification (a 10-day notice). However, SMCRA limits OSM's authority during such an inspection. OSM may issue *only* a cessation order (CO) and it may issue that order *only* when the alleged violation of the act creates an "imminent danger to the health or safety of the public, or is causing, or can reasonably be expected to cause significant, imminent environmental harm" (30 U.S.C. section 1271[a][2]). Notwithstanding that statutory limitation, the regulations permit OSM to issue a notice of violation (NOV) based on the federal inspection, regardless of whether the violation is an imminent danger, if the state fails to take appropriate action within 10 days, or fails to show good cause why it has not done so (30 C.F.R. section 843.12[a][2]). OSM's authority to issue federal NOVs in primacy states has been challenged in National Coal Association, et al. versus Lujan (C.A. nos. 87–2076–JGP and 88–2416–JGP [consolidated]), pending before the United States District Court for the District of Columbia.

SMCRA itself authorizes OSM to issue a NOV only when OSM finds a nonimminent danger violation during a federal inspection pursuant to a federal program, a federal lands program, the interim program, or the substitution of federal enforcement where the state has failed to enforce its state program (30 U.S.C. section 1271[a][3]). OSM is *not* authorized to issue a NOV in a primacy state where the approved state program is in full force and effect.

SMCRA's legislative history clearly and strongly supports a statutory construction that OSM has no authority to issue directly NOVs in its oversight of the states' administrative action pursuant to sections 504(b) and 521(b) of the act and part 733 of the agency's regulations. Contrary to OSM's conclusions in 1979 that Congress left a "statutory void" (44 *Federal Register* 15302 [March 13, 1979]), the legislative history discloses that the absence of direct federal NOV authority under the permanent program was clearly intended by both houses (see H.R. Rep. No. 218, 95th Cong. 1st Sess. [1977]; S. Rep. No. 128, 95th Cong., 1st Sess. [1977]; H.R. Rep. No. 493, 75th Cong., 1st Sess. [1977]).

A review of the manner in which OSM has employed the NOV as an oversight tool discloses that instead of filling some perceived statutory void, the agency has created for itself a regulatory loophole to circumvent established procedures for resolving oversight problems with the states. OSM originally invoked NOV authority to fill the so-called gap because it anticipated that states would ignore violations of the law. However, today, the typical federal NOV arises from OSM's disagreement with a state's interpretation of its program, the manner in which a state compels abatement of violations, alleged discrepancies in state-issued permits, and even deficiencies in state programs approved by OSM. In each instance, the statute and regulations establish procedures for an appropriate OSM response to resolve the matter: program amendments, 30 C.F.R. 732, permit revisions, 30 C.F.R. 774, and preemption of state law, 30 C.F.R. 733.

OSM's reflexive use of oversight NOVs to address the states' implementation of their programs has rendered the full range of oversight tools superfluous. Again, from a policy perspective, OSM would be better advised to use the established procedures for program amendments and permit revisions to resolve program deficiencies, interpretational disputes, and permit discrepancies. To the extent a state response to a 10-day notice discloses a view of a program or permit requirement different than that held by OSM, the program amendment or permit revision processes reach the issue at a broader program level. Moreover, the underlying issues in controversy between OSM and the state deserve meaningful consideration through a rule-making proceeding with broader public participation than that which occurs in an isolated enforcement proceeding.

In passing SMCRA, Congress anticipated that a state might fail to properly enforce its state program. However, the mechanisms it provided to deal with such occurrences did *not* include blanket OSM authority to issue overriding federal NOVs in primacy states. Instead, the act provides for OSM to take over enforcement of the state program (either wholly or partially) when the state has failed to enforce it. The act requires that, when OSM has reason to believe that the state is failing to enforce effectively all or part of any state program, OSM must give public notice and notice to the affected state, and then it must hold a hearing to determine whether the "State has not adequately demonstrated its capability and intent to enforce such State program" (30 U.S.C. 1271[b]). If, after the hearing, OSM finds that the state either cannot or will not enforce the state program, OSM takes over administration of the state program, including permitting, inspections, and enforcement action (30 U.S.C. section 1271[b]; 30 C.F.R. section 733.12).

The substitution of federal enforcement for any part of a state program or the promulgation of a federal program remain OSM's ultimate power over lax state enforcement (30 U.S.C. 1271 [b]; see also, In re Permanent Surface Mining Regulation Litigation [en banc], 653 F. 2d at 520; Clinchfield Coal Co. v. Hodel, Memo Op. at 2). The statute sets forth a two-tiered test for substituting federal enforcement or for withdrawing program approval: (1) violations exist from the failure of the state to enforce all or any part of its program effectively, and (2) the state has not adequately demonstrated its capability and intent to enforce its program (30 U.S.C. 1271[b]; see also, 30 C.F.R. 733.12[e]).

While the actual substitution of federal enforcement or a federal program is clearly intended as a last resort, OSM has ignored the initiation of such proceedings as an oversight tool to address widespread enforcement disparities. Nothing mandates that OSM withdraw approval or substitute enforcement after it has decided to hold a hearing pursuant to section 521(b). In fact, the implementing regulations (30 C.F.R. part 733) afford ample opportunity to resolve the issues with the state short of substituting either federal enforcement or a federal program. For example, the regulations require that OSM provide the state with notice and time to respond to allegations. The state may also request a conference to resolve the dispute. If the problem persists, then OSM holds a public hearing to gather further information and comments. The results of the hearing and investigation formulate the basis of OSM's findings that the state has corrected the problem, or that the state has not demonstrated the capability and intent to do so. Only if OSM makes the latter finding do the sanctions of substituted federal enforcement and/or withdrawal of program approval follow.

The implications for the congressionally contemplated scheme of primacy, should OSM prevail in its use of federal NOVs, are significant and potentially debilitating. In addition to making a mockery of the exclusive jurisdiction vested in the states by SMCRA, the use of federal NOVs breeds dual enforcement of the act and results in conflicting, confusing, and counterproductive signals being sent to the regulated industry. It also fosters excessive litigation over issues of jurisdiction and of regulatory and policy interpretation. In the majority of cases it places the coal operator in the middle of disputes between two regulatory authorities. In the final analysis, the states find that they have little in the way of true primacy or exclusive regulatory authority.

The states have gone so far as to suggest that OSM establish a mechanism to resolve those issues where differences of interpretation or jurisdiction arise between the two arms of government. However, that suggestion has fallen on deaf ears, presumably because OSM finds it easier to simply directly enforce against the states when a question of authority is raised. Although cleaner and less complicated, such an approach is unfortunately without statutory authority and flies in the face of congressional intent.

FUTURE FEDERALISM CONCERNS

State governments are, as Thomas Jefferson claimed, in a better position to serve people because they are closer to the people.

Environmental protection and resource management call for a stabilizing federal presence, but the federal government must guard against fostering well-intentioned programs that produce costly activity without progress. Problems are inevitable with federal legislation that paints the entire nation with the same broad stroke. Americans live in a land of diverse environmental conditions and problems. Federal regulation of our environment must reflect the diversity of this country's many regions. Efforts to achieve and sustain a cleaner world require a balanced partnership between the states and the federal government, an arrangement that recognizes and builds upon the relative strengths of the partners.

Two principles emerge as we chart the course of environmental protection in general. First, there is clearly a role for a significant federal presence. The second, and balancing, principle is that environmental programs must involve the states in a substantive role as environmental action is assimilated into the local environmental and social order. These principles are the basis for sound federalism in which each level of government works within its proper sphere and authority to provide service to its citizens. Healthy tension between the states and the federal government works to moderate excessive action or to prevent failure to act.

For their part, the states shoulder the primary responsibility for planning, designing, implementing, and enforcing programs to achieve federal and state goals and standards. This involves exercising discretion in the design and operation of environmental programs, as long as program goals are achieved. It also involves the right to establish standards more stringent than federal minimums, in accordance with the fundamental obligation of states to protect the health and welfare of their citizens.

The focus of federal efforts in environmental protection should be on those several things only the federal government can do effectively. First, federal statutes must address pressing environmental problems, including interstate disputes, in order to encourage responsible action from the governments involved. Second, there is a logical and scientific basis for the federal government to set national minimum standards to protect public health, the general welfare, and the environment. And third, the federal government should provide state and local environmental agencies with technical, financial, and managerial assistance and with performance-based oversight. In the late 1970s, when a cleaner world became a national priority, and then a national commitment, states welcomed such federal assistance. That assistance built, and continues to sustain, a myriad of state programs.

When Congress enacted the surface mining act, the states believed its intent was to ensure uniform standards throughout the nation, preventing any one state from having a competitive advantage by virtue of less stringent state regulatory requirements. Many states supported SMCRA, believing that such federal legislation, in concert with their own recent regulatory efforts, would help create a fair and effective system.

What the states anticipated and what they got from the federal legislation were not the same. The Department of the Interior insisted that each state submit a proposed regulatory program that was little more than a carbon copy of the program embodied in the federal legislation and rules. Small differences that reflected local preferences or unique state conditions were discouraged. Requirements that the states argued were "as effective as" the federal program were greeted with suspicion and mistrust by federal regulators. State flexibility was not what those on the Potomac wanted to see in a program. Repeatedly, federal bureaucrats reacted either by denying state requests or by requiring seemingly endless justifications.

Once state surface mining programs were approved, state-federal relations were strained even more as the Department of the Interior insisted upon lengthy written descriptions of the decision-making process for every mine-plan decision. In many states, the ability to work out problems on the ground with the operator has been impaired because of the extensive time required to satisfy Department of the Interior overseers. Clearly, there have been twin problems of turf and mistrust.

Several tests lurk on the horizon for the state-federal partnership. Perhaps the most telling will be the implementation of a recently negotiated settlement concerning OSM's Applicant Violator System (AVS). The AVS is a nationwide computer system aimed at blocking permits to coal operators with outstanding violations or civil penalties under SMCRA. Under the recent settlement, the states are called upon to screen applicants and assist in building the AVS data base, to determine and investigate links between permit applicants and outstanding violations and violators, and, most importantly, to act on recommendations provided by the AVS concerning the decision to issue or deny a permit based on data in the system.

Once state regulatory programs are approved, the states exercise exclusive jurisdiction under SMCRA, and they are responsible for issuing permits to authorize legal mining and reclamation activities pursuant to the provisions of the state program. AVS relies on the permitting function to determine links between permit applicants and outstanding violations or violators. Thus, without cooperation from the states (as sole issuers of permits), the AVS could not operate (for a more detailed history of the AVS and the problems of gross abuse which led to its eventual establishment, see Conrad 1987).

The AVS has the potential to undermine the states' permitting authority, particularly under the terms of the recent settlement, since it incorporates a federal enforcement scheme where states fail to cooperate. Although the state is the *dejure* permitting authority, the Applicant Violator System makes OSM the permitting authority in fact because it requires that all permitting decisions be based on the OSM-maintained system that the states cannot control. In essence, the AVS relegates the primacy state to the position of a rubber stamp of OSM.

Another testing ground is that of state program amendments. Again, recognizing the value and appropriateness of states being more in touch with local conditions and with geographic, climatic, and other environmental differences from region to region, Congress provided for the development and approval of individual state programs that were to be consistent with the federal regulations. Although the window of opportunity for states to tailor their programs to the needs and conditions prevailing within their respective borders has been likened to a "one way mirror" rather than a true "state window" (In re Permanent Surface Mining Regulation Litigation, 653 F.2d 514,532 [Tamm, J., dissenting] [D.C. Cir.

1981] [*en banc*], *vacating*, 14 E.R.C. 1813 [D.C. Cir. 1980]), states have persevered in their efforts to fashion amendments to their primacy programs that reflect current conditions and changed circumstances. Many of these efforts have been in response to extensive litigation on OSM's regulations, whereby numerous regulations have been struck down, suspended, remanded, withdrawn, reinstated, modified, and repromulgated. The states have had to move in response to numerous fits and starts as the federal regulators have changed their minds or suffered defeats in the courts.

Most recently, OSM disapproved program amendments by Virginia and Kentucky intended to provide alternative permitting and performance standards for remining (55 *Federal Register* 13158, April 9, 1990; 55 *Federal Register* 2240, January 23, 1990). The intent of the amendments was to provide a more flexible regulatory framework to encourage reclamation of abandoned mine lands through remining. The disapproval came almost four years after Kentucky submitted its amendment, and more than two years after Virginia submitted its amendment.

OSM's general reason for disapproving the remining packages was the absence of any federal counterpart. While OSM has, for several years, studied the possibility of establishing a comprehensive set of remining standards, it has not yet formally proposed any such standards. Another reason OSM offered for disapproving the remining amendments was the lack of "clarity of intent" in several provisions. In other words, OSM was uncomfortable with the discretion the states would have in the practical application of these rules. Finally, while OSM officials did find some of the provisions "no less effective" than federal standards, they disapproved the complete package since, in their view, the approvable provisions could not stand alone from the whole package. It appears that OSM will continue to thwart state initiatives in this area until the federal agency decides to take the lead.

OSM has recently developed guidelines for approving state program amendments—guidelines that are intended to expedite and facilitate what, to date, has been an unduly burdensome, cumbersome, and time-consuming process. The critical issue, however, is the federal government's willingness to consider and accept amendments to state programs that are soundly based on local or regional needs or circumstances, regardless of their mirror image likeness to federal counterpart rules. Only in this way can the federal system truly work. It must remain dynamic and flexible, thus encouraging and facilitating change. The federal system works because it provides opportunities for experimentation and innovation. It works because it allows for diversity among the states and because, by preserving government close to the people, it assures greater responsiveness and accountability.

Perhaps the area with the most potential in terms of federal-state relations is oversight. It serves as the laboratory and testing ground for the concept of cooperative federalism in terms of communication and problem identification and resolution between the state and federal governments. Oversight depends upon a clear understanding of and appreciation for the respective roles of these governments, pursuant to SMCRA. But mostly, it calls for an appropriate attitude on the part of those who must deal with each other on a day-to-day basis to accomplish mutual purposes and objectives.

The importance of attitude on the ultimate success of an oversight program simply cannot be overlooked. A predisposition by federal decision makers or policy makers toward a heavy national influence will affect their ability to reflect open-mindedly and even-handedly on matters of primacy. Similarly, an overly aggressive attitude about states' rights, to the exclusion of an appropriate oversight role by the federal government, will preclude a successful cooperative effort. To strike an effective balance between state and federal government is the goal of the federal system, and to approach such a balance with the right attitude is the key to its accomplishment.

Unfortunately, in the eyes of many in the federal government, the states are little more than interest groups that must lobby and litigate to protect the interests of their citizens. Nor has Congress shown much interest in correcting the growing imbalance in the federal system. Congress, the courts, and the executive branch have continued to expand federal powers. Two startling indicators of the erosion of state and local authority stand out: (1) in our nation's two hundred-year history, more than half of all federal statutes that preempt state and local authority have been enacted during the last 20 years (about 190 statutes out of some 354); and (2) more than 35 percent of all United States Supreme Court rulings that declare a state or local act to be unconstitutional have been issued during the past 25 years (Nething 1990).

The fact is, however, that states are able, willing, and eager to govern effectively. As one noted expert in the field tells it, "If any lessons are to be learned from the organization of effective institutions today, it is that non-centralization, networking and diversity work better than hierarchies. The kind of hierarchy that has developed in the federal system—one in which state and local governments are viewed as lower *levels*—undermines the capacity for innovation, adaptation and flexibility originally built into the federal system" (Nething 1990).

The states continue to see signs of progress that the federal government will come around and begin to treat them as true partners. On February 16, 1990, President George Bush signed a Memorandum on Federalism that reaffirms an earlier Executive Order (No. 12612) signed by former President Ronald Reagan. President Reagan's executive order was hailed by the states as a clear and comprehensive articulation of the appropriate relationship that must be in place between the two governments to accomplish the responsibilities of government. In his memorandum, President Bush emphasized several aspects of the earlier executive order, including the following:

1. In most areas of governmental concern, the states uniquely possess the constitutional authority, the resources, and the competence to discern the sentiments of the people and to govern accordingly.
2. With respect to national policies administered by the states, the national government should grant the states the maximum administrative discretion possible.
3. When undertaking to formulate and implement policies that have federalism implications, federal executive departments and agencies should (1) encourage states to develop their own policies to achieve program objectives and to work with appropriate officials in other states; (2) refrain, to the maximum extent possible, from establishing uniform national standards for programs and, when possible, defer to the states to establish standards; and (3) when national standards are required, consult with appropriate officials and organizations representing the states in developing those standards.

President Bush also stressed the responsibility of each federal agency to prepare a "federalism assessment." This action is required when any proposed federal policy has sufficient potential implications for the states as to require an evaluation of the potential costs or burdens on state governments. Each agency is also to appoint an official responsible for ensuring that the agency will implement the order.

Of course, the proof is in the application. The effectiveness of any executive order on federalism will be tested in the day-to-day attitudes of those in the federal agencies who must work with the states and in the federal policies and rules that impact the states. The states will have to remain on their guard, as suggested by the words of former Governor John Sununu, who recently said:

With the 10th Amendment [the founding fathers] thought they were saying that there are a great number of powers reserved for the states and the only ones we are not reserving for the states we are carefully defining. But this constitution has been interpreted so broadly and generally in the wrong direction that there are many folks now—in fact virtually all constitutional lawyers—who would suggest that the 10th Amendment in fact is moot and has no credibility. (Sununu 1990)

CONCLUSION

Based on the lessons learned and the hard tests ahead, the time has come for all affected parties under SMCRA—state and federal governments, environmentalists, citizens, industry—to ask and to answer the question: Are the states to function in the role of primary regulators of SMCRA and, if so, what does that role mean?

At the time of SMCRA's enactment, the answer was clearly "yes—the states should have a significant role as the primary government regulatory authority under the act." This answer, however, was somewhat speculative, as it occurred without the benefit of experience. Now, after almost 10 years of experience under primacy, this answer must be reevaluated from everyone's perspective; the question must be asked again and answered with confidence and with a full understanding of its implications. If adjustments to the regulatory regime are required, they should be placed on the table and discussed and resolved. In the end, the regulatory program can operate effectively and the purposes and goals intended by Congress will be accomplished.

NOTES

1. Consideration of proposed legislation that led ultimately to the act began in 1968 (Surface Mining Reclamation: Hearings Before the Senate Committee on Interior and Insular Affairs, 90th Cong., 2d. Sess. [1968]). Further hearings were held in both houses in 1971, and the House passed a predecessor version of the act in 1972 (H.R. 6482, 118 Cong. Rec. 35.031, 35,055–56 [1972]). Congress actually passed earlier versions of the act in December 1974 and in May 1975, but both were vetoed by President Ford. By the time President Carter signed the act into law on August 3, 1977, SMCRA had thus generated an extended body of hearings, committee reports, conference committee reports, and legislative debates. The legislative history is summarized in Note, *West Virginia Law Review* 81: 775 (1979). Legislative developments prior to the 95th Congress, which passed the act, are summarized also in Senate Report No. 95–128, 95th Cong., 1st Sess. 59–61 (1977), and House of Representatives Report No. 95–218, 95th Cong., 1st Sess. 140–41 (1977) (reprinted in *U.S. Code Cong. and Admin. News* [1977] 593).

2. For further information regarding the congressional and judicial activity that has attended SMCRA's enactment and implementation, see Abrams 1982;

Barry 1981; Brion 1981; Edgmon and Menzel 1981; McElfish and Beier 1990; McGraw 1980; National Coal Issue 1979; Regulation Without Just Compensation 1981; and Rasnic 1982.

REFERENCES

Abrams, Dennis. 1982. "The Federal Surface Mining Control and Reclamation Act of 1977—First to Survive a Direct Tenth Amendment Attack." *West Virginia Law Review* 84: 1069.

Barry, Hamlet J. 1981. "The Surface Mining Control and Reclamation Act of 1977 and the Office of Surface Mining: Movable Targets or Immovable Objects?" *Rocky Mountain Mineral Law Institute* 4: 117.

Brion, Denis J. 1981. "Federal Regulation of Surface Mining in Virginia." *Virginia Law Review* 67: 329.

Conrad, Gregory E. 1987. "The Applicant Violator System Under the Surface Mining Control and Reclamation Act of 1977: Constitutional Concerns." *Eastern Mineral Law Institute* 8: ch. 10.

"A Constitutional Challenge to the Surface Mining Control and Reclamation Act." 1981. *Public Lands Law Review* 2: 138.

Edgmon, Terry D., and Donald C. Menzel. 1981. "Regulation of Coal Surface Mining in a Federal System." *Natural Resources Journal* 21: 245.

Hawkins, Jr., Robert B. 1990. "The Dramatic Rise of Federal Supremacy." *Journal of State Government* 63: 10.

McElfish, James, and Ann Beier. 1990. "Environmental Regulation of Coal Mining: SMCRA's Second Decade." Environmental Law Institute.

McGraw, Marcus. 1980. "Environmental Regulation of Coal Mining—New Directions." *Eastern Mineral Law Institute* 4: 117.

Macleod, John A., Thomas C. Means, and Susan E. Chetlin. 1988. "State Primacy Under the Surface Mining Act: The First Decade." *Eastern Mineral Law Institute* 9: ch. 5.

"National Coal Issue—A Symposium on the Surface Mining Control and Reclamation Act of 1977." 1979. *West Virginia Law Review* 81: 553.

Nething, David E. 1990. "States Must Regain Their Powers." *Journal of State Government* 63: 6.

Rasnic, Carol C. 1982. "The Conflict Between the Environmentalists and the Surface Mining Industry: Federal Legislation and Litigation." *Northern Kentucky Law Review* 9: 427.

"Regulation Without Just Compensation: A Political Process-Based Takings Analysis of the Surface Mining Act." 1981. *Georgetown Law Journal* 69: 1083.

Sununu, John. 1990. "The Tough Battle Ahead." *Journal of State Government* 63: 25.

III Courts and Implementation

7 Impacts of Federal Court Decisions on the Administration and Implementation of the Surface Mining Control and Reclamation Act

Robert E. Beck

The Surface Mining Control and Reclamation Act (SMCRA) was enacted to regulate various aspects of coal mining and coal mine reclamation, and it sets up a particular regulatory scheme for doing so. This scheme includes creating the Office of Surface Mining Reclamation and Enforcement (OSMRE), a federal regulatory agency within the Department of the Interior, and providing the opportunity for states to assume the primary role in administering and enforcing SMCRA.

This chapter deals with federal judicial review of actions or inactions of OSMRE and discusses the impacts court decisions have or can have on the administration and implementation of SMCRA. Before considering the role of court decisions in relation to SMCRA, it is necessary to briefly examine three elements: the general federal administrative process; the role of the federal courts generally in relation to that administrative process; and the issue of who generally may bring matters involving that administrative process to court for review.

THE ADMINISTRATIVE PROCESS

Federal agencies, located generally within the executive branch of the federal government, exercise quasi-legislative and quasi-judicial powers as well as carry out executive functions (Gellhorn and Levin 1990; Schwartz 1984). Exercising quasi-legislative power, agencies promulgate regulations to flesh out the federal legislation that sets up the basic program they are to carry out (rule making); using quasi-judicial power, they enforce the law at the primary level through means such as issuing citations, revoking licenses, and imposing civil penalties (adjudication).

In addition, they issue licenses or permits. Because issuing licenses and permits may involve the frequent exercise of both powers, it is difficult to categorize that process into one or the other camp. However, whether the agency is engaged in rule making, in adjudication, or in both, it is interpreting the law, finding legislative or adjudicative facts, and using discretion to apply the law to the facts.

COURT REVIEW OF THE FEDERAL ADMINISTRATIVE PROCESS IN GENERAL

As Gellhorn and Levin point out, legal scholars recognize three purposes for court review of the administrative process: (1) to assure that the agency acts in accord with the will of the political branches, as expressed in the enabling legislation; (2) to foster reasoned decision making by requiring supporting facts and rational explanations; and (3) to legitimize bureaucratic regulation by way of an independent decision about the validity of the administrative determination (Gellhorn and Levin 1990, 73–74).

In reviewing interpretation of the law, two relevant questions arise under the Administrative Procedures Act (APA): whether the Constitution of the United States has been violated (5 U.S.C. section 706[2][b]), and whether the agency's statutory authority has been exceeded (5 U.S.C. section 706[2][c]). As to fact finding, the APA provides three types of approaches to or standards for judicial review: (1) the substantial evidence rule (5 U.S.C. section 706[2][e]); (2) the arbitrary and capricious standard (5 U.S.C. section 706[2][a]); and (3) *de novo* review (5 U.S.C. section 706[2][f]), which is largely nonexistent in the federal administrative law scheme.

As to agency action in applying the law to specific facts, the general administrative law scheme recognizes that the agency is to have considerable discretion. This is the area in which the agency is to develop its expertise. Gellhorn and Levin (1990) claim that "the court has two tasks: To ensure . . . that the agency does not exceed the outer bounds of the delegation; and . . . that the delegated power is exercised in a rational fashion" (77).

The APA provides for judicial review of agency procedure allowing courts to invalidate proceedings taken "without observance of procedure required by law" (5 U.S.C. section 706[2][d]). Finally it also specifies that if an agency has failed to take an action required by Congress or has unreasonably delayed that action, the federal courts have authority to compel the agency to act (5 U.S.C. section 706[1]).

As long as the courts exercise meaningful review, the mere availability of judicial review can serve as a restraint on unbounded action or inaction by an agency. However, the doctrine of standing is central to the availability of judicial review.

Standing

The issue of who may take an agency action to court for review is dealt with under the concept of standing. There never has been any real problem with the person being regulated having standing to go to court to challenge a regulation, a citation, a civil penalty, or any other agency action that affects the party's economic interest. However, it has not always been clear to what extent third parties, that is, someone other than the agency and those being regulated, may go to court. Many persons seem to have assumed that because Congress has provided a role in the administrative process for the public generally, members of the public may take a matter to court. However, it is not that simple. Plaintiffs have to show three things: (1) that plaintiff "personally has suffered some actual or threatened injury as a result of the putatively illegal conduct of the defendant"; (2) "that the injury can be fairly traced to the challenged action"; and (3) that the injury "is likely to be redressed by a favorable decision" (Valley Forge College v. Americans United, 454 U.S. 464, 472 [1982]). This test makes standing a question of degree and the borderline vague.

Failure to meet two additional doctrinal requirements, ripeness and exhaustion, also may prevent a court from reviewing agency action or inaction, but they are beyond the scope of this chapter.

SMCRA

Under SMCRA, the secretary of the interior performs four basic functions: (1) promulgating regulations, (2) approving state primacy programs, (3) overseeing state administration and enforcement once state primacy is achieved, and (4) directly administering and enforcing SMCRA when state primacy does not exist. Although the day-to-day federal activity in implementing SMCRA occurs at the Office of Surface Mining Reclamation and Enforcement, the secretary of the interior is the final decision maker. The secretary has delegated some administrative review responsibilities to the Department of the Interior's Office of Hearing and Appeals. (For a thorough review of administrative review, see Irwin 1988.) The detail of the administrative review process is beyond the scope of this chapter; however, the role of judicial review in each of the four functional areas will be discussed, perhaps giving some impression about whether the function involved affects the impacts of judicial review. In addition, the doctrine of standing as applied in SMCRA cases will be discussed.

Because the Supreme Court of the United States has already found those provisions of SMCRA most likely to receive a direct constitutional challenge to be facially constitutional,[1] this chapter focuses on judicial review of administrative action.

SMCRA contains explicit judicial review standards so that generally it is unnecessary to refer to the standards in the Administrative Procedures Act (5 U.S.C. sections 551–706). However, courts will look to decisions interpreting similar language in the APA for guidance in interpreting the SMCRA review language.

Legislative Rule Making

In the process of rule making, an administrative agency both follows procedures and makes substantive decisions about what the statute permits, what should be regulated, and what should be required of the parties that are subject to regulation. According to SMCRA, regulations approved by the secretary are to be upheld, "unless the court concludes that such action is arbitrary, capricious, or otherwise inconsistent with law" (section 526[a][1]; 30 U.S.C. section 1276[a][1]). The net result of this language is that a court may have to make three basic inquiries in reviewing the secretary's regulations: (1) whether the secretary followed proper procedures; (2) whether the challenged regulation is contrary to law; and (3) whether the challenged regulation is arbitrary and capricious.

The United States Supreme Court has concluded that as to rule making, the APA establishes generally the maximum procedural requirements Congress was willing to have the courts impose upon agencies (Vermont Yankee Nuclear Power Corp. v. NRDC, 435 U.S. 519, 524 [1978]). However, "agencies are free to grant additional procedural rights in the exercise of their discretion, but reviewing courts are generally not free to impose them if the agencies have not chosen to grant them" (Vermont Yankee Nuclear Power Corp. v. NRDC, 435 U.S. 519, 524 [1978]). The APA section that the court referred to provides that "general notice of proposed rule making shall be published in the Federal Register," and "after notice . . . the agency shall give interested persons an opportunity to participate in the rule making through submission of written data, views, or arguments with or without opportunity for oral presentation" (5 U.S.C. section 553[b] and [c]). The notice is to include (1) "the time, place, and nature of public rule making proceedings"; (2) "reference to the legal authority under which the rule is proposed"; and (3) "either the terms or substance of the proposed rule or a description of the subjects and issues involved" (5 U.S.C. section 553[b]). The notice and opportunity for comment requirements are written in SMCRA. However, SMCRA goes beyond the APA and requires at least one public hearing (section 501[a] and [b]; 30 U.S.C. section 1251[a] and [b]).

A useful illustration of a procedural problem in the context of SMCRA involves the duty of a mine operator to repair damage caused by subsidence. The secretary's permanent program regulations promulgated in 1979 provided that an underground mine operator who caused subsidence damage had to repair or restore both land and structures damaged by the subsidence (30 C.F.R. section 817.124 [1979]). In an attack by industry, the federal district court upheld the regulation (In re Permanent Surface Mining Regulation Litigation, 14 E.R.C. 1083, 1108 [D.D.C. 1980] [round I]). This decision was appealed, but because OSMRE decided to undertake a comprehensive review of the 1979 regulations, the appeal did not go forward. The revised permanent program regulations promulgated in 1983 contained provisions on subsidence that were different from the 1979 regulations (48 *Federal Register* 24,647 [1983]). The bulk of the revised subsidence regulations that were challenged were approved by the federal district court in 1984 (In re Permanent Surface Mining Regulation Litigation, 21 E.R.C. 1724 [D.D.C. 1984] [round II]).

The revised subsidence regulations, however, qualified the mine operator's duty to repair and restore the surface, depending on whether the damage was to land or to structures (30 C.F.R. section 817.121[c][1] [1983]). If the surface damage was to land, the underground mine operator had to repair or restore the damage to the same extent as under the 1979 regulations. This requirement was approved by the district court (round II at 1727–32). If the damage was to a structure, the operator would have no duty to repair or restore the structure unless state law required that the structure be repaired or restored. When the secretary first proposed the revised rules, the proposal did not contain this division between land and structures; the change appeared only when the final rules were promulgated. According to the district court, this final rule was "a radical change from the proposed rule," involving "a complete reversal of policy from the proposed regulation" (round II, at 1730). The net effect was that the parties to the rule-making proceedings had been deprived of any meaningful opportunity to comment on the change in operator responsibility contained in the final rule. Consequently, the district court remanded the rule to the secretary for further action.

In February 1985, the secretary suspended the language, "to the extent required under state law," leaving the rest of the rule: damage to structures must be repaired or the structure restored, as promulgated (50 *Federal Register* 7,276 [1985]). In July 1985, the secretary reproposed the rule as it had been promulgated in 1983 (50 *Federal Register* 27,910 [1985]), with language that read, "to the extent required under state law." In February 1987, the secretary promulgated the final rule, basically as it had been promulgated in 1983, and thus separated the duty to repair or restore as between land and structures (52 *Federal Register* 4860 [1987]). In April 1987, the National Wildlife Federation brought suit, challenging the final regulation (National Wildlife Federation v. Hodel, "Complaint for Declaratory and Injunctive Relief, and Petition for Review" [April 15, 1987]). In February 1990, the district court found the language, "to the extent required under state law," to be inconsistent with SMCRA and ordered the secretary to remove it (National Wildlife Federation v. Lujan, 733 F. Supp. 419 [D.D.C. 1990]). In March 1991 the court of appeals reversed, finding that Congress had expressed no clear intent on the matter and that the secretary's interpretation was a reasonable interpretation (National Wildlife Federation v. Lujan, 928 F.2d 453 [D.C. Cir. 1991]). Because the court also rejected the industry argument that SMCRA precluded the secretary from requiring mine operators to compensate for damaged structures and because the secretary had started to implement the district court's decision, it is uncertain what the secretary will do now. He may decide to accept the court of appeals reversal or he may seek to revise the rule again to require compensation for damage to structures.

The subsidence regulation case illustrates that the process can be a long one. The first step was taken when the regulations were promulgated in 1979, and the process still is not over despite the decision of the appellate court. Of course, the more than 10-year delay in settling the subsidence regulation has resulted in part from the intervention of a new administration—one with different political views.

The subsidence regulation case also illustrates what can happen when procedural inadequacies are the bases for overturning agency action. The court dealt initially with the agency's failure to provide adequate notice of the proposed rule and remanded the action back to the agency; the agency corrected the procedural error by publishing notice of the proposed rule change and came out with the same substantive revised rule as before. However, when the court reviewed the substance of the agency action, it overturned the action again but was then reversed by the appellate court. This leads us directly to a discussion of the next aspect of judicial review—whether the challenged regulation is contrary to law.

Two "contrary to law" aspects must be considered with respect to SMCRA. First, is the regulation within the area delegated to the agency for regulation? And second, has the agency correctly interpreted the law?

One of the 1979 regulations that industry challenged required permit applicants to submit considerably more data than was required by the 17 items Congress specified in SMCRA (section 507[b][1]-[17]; 30 U.S.C. section 1257[b][1]-[17]). The challenge turned into a review of the general scope of the secretary of the interior's rule-making power under SMCRA. While SMCRA contained two general provisions on rule making (SMCRA section 201[c][2]; 30 U.S.C. section 1211[c][2]; SMCRA section 501[b]; 30 U.S.C. section 1251[b]), industry argued that in those provisions Congress intended to limit the secretary of the interior to housekeeping-type regulations, such as how to allocate funds and who was to do what within the department. Industry argued further that Congress intended to allow substantive regulations only where SMCRA specifically said that the secretary could promulgate substantive regulations. Industry noted 21 specific grants of rule-making power. In 1981, the federal court of appeals in the District of Columbia decided on a five-to-four vote that Congress intended the scope of the secretary of the interior's rule-making power to be broad (In re Permanent Surface Mining Regulation Litigation, 653 F.2d 514 [D.C. Cir. 1981] [en banc], *cert. denied* 454 U.S. 822 [1981]).

While the regulation seemed to focus narrowly on the data to be provided with the mine permit application, the regulation had broad-based implications for the administration of SMCRA; the secretary of the interior intended to use the additional data in considering state primacy requests and in conducting oversight inspections and proceedings. Primarily, however, the secretary hoped the requirements for additional data would ensure that the states acquired sufficient information to enforce the SMCRA requirements and thus reduce the number of instances in which the secretary might have to directly enforce SMCRA or resume general administration. Without the ability to require this data, the secretary would have been limited to inspecting more rigorously, to trying to influence states to require the data by issuing interpretations, or to requesting the states to get the data.

Ultimately, the most comprehensive review of individual regulations promulgated under SMCRA to come before the federal court of appeals for the District of Columbia circuit occurred in 1988 in National Wildlife Federation v. Hodel (839 F.2d 694 [D.C. Cir. 1988]); while there have been many phases to the challenge of the SMCRA regulations, with separate batches considered at different times, most of the discussion in this chapter relies on the 1988 opinion for illustration. The court reviewed 28 regulations that had been reviewed by the federal district court. Industry challenged 5 of the 28 regulations, while environmental groups challenged the remaining 23. Both the district court and the appellate court rejected all 5 of the industry challenges. The district court ruled in favor of 15 of the environmental group challenges; the appellate court agreed in only 8 of those 15 challenges. The appellate court agreed with the district court on all 7 of its rulings against the claims of the environmental groups. Taking stock of numbers is useful only in a limited sense, for the numbers do not deal with the complexity of the issue involved. However, they can reinforce analyses of the overall approach and the perception of overall problems.

Obviously, 28 regulations challenged out of all those promulgated under the SMCRA are not very many. There are 408 pages of general SMCRA regulations and 233 pages of state program regulations in the 1990 edition of volume 30 of the *Code of Federal Regulations*. However, over the course of the challenges to the regulations, many district court decisions against the secretary of the interior were not appealed. In part at least this resulted from two administrations being involved. A new administration may have agreed with the court's decision overturning a regulation, and thereby it may have avoided the need to change the regulation on its own motion. However, the split in the results in the other cases before the federal district court is not appreciably different. For example, the 1990 rounds have elicited a similar score, with 19 for the secretary, 7 for challengers, and 3 mooted (National Wildlife Federation v. Lujan, 733 F. Supp. 419 [D.D.C. 1990]; 31 E.R.C. 1617 [D.D.C. 1990]; 31 E.R.C. 2034 [D.D.C. 1990]; 31 E.R.C. 2069 [D.D.C. 1990]). (For a review of other rounds, see Case 1979. For a listing of all the court review cases, see National Wildlife Federation v. Lujan, 733 F. Supp. 419, 421–22 n.2 [D.D.C. 1990].)

Two of the regulations involved in the 1988 review are useful to illustrate the impact of a prior decision. Both regulations were successors to regulations previously rejected by the district court but now upheld in their new form. The successor regulations took into account what the district court had found wrong with the predecessor regulations.

One of the two involves instances when mine operators have to replace damaged water supplies (section 717[b]; 30 U.S.C. section 1307[b]). The 1979 regulations promulgated by Secretary of the Interior Cecil Andrus required all damaged water supplies to be replaced, regardless of whether surface mines or underground mines were involved (44 *Federal Register* 15,430 [1979]). In 1980, the district court, pursuant to an industry challenge, ruled that SMCRA required only *surface* mine operators to replace damaged water supplies. (Permanent Surface Mining Regulation Litigation I [round II], 19 E.R.C. 1477, 1495 [D.D.C. 1980]). When Secretary of the Interior Watt promulgated the revised regulations in 1983, the revised regulations did not require underground mine operators to replace damaged water supplies (48 *Federal Register* 43,956 [1983]). The National Wildlife Federation challenged the new regulation. The district court reaffirmed its 1980 interpretation of SMCRA and upheld the new regulation (Permanent Surface Mining Regulation Litigation II [round III], 620 F. Supp. 1519, 1532–33 [D.D.C. 1985]). The appellate court agreed (839 F.2d at 753–54).

The other of the two regulations involves air quality. Secretary Andrus, in the 1979 regulations, placed limitations on fugitive dust from coal mine sites (44 *Federal Register* 15,050 [1979]). Fugitive dust was defined as "particulate matter not emitted from a duct or stack which becomes airborne due to the forces of wind or surface coal mining operations" (44 *Federal Register* 15,318 [1979]). In 1980, the district court held that SMCRA did not authorize regulation of fugitive dust itself, but only in the context of erosion (Permanent Surface Mining Regulation Litigation I [round II], 19 E.R.C. 1477, 1491–92 [D.D.C. 1980]). Secretary Watt's revised regulations in 1983 limited air pollution prevention to that arising from erosion at the mine site (48 *Federal Register* 1160 [1983]). The National Wildlife Federation challenged this limitation. The district court reaffirmed its 1980 interpretation of SMCRA and upheld the revised regulation (Permanent Surface Mining Regulation Litigation II [round I], 21 E.R.C. 1193, 1206–07 [D.D.C. 1984]). The appellate court agreed (839 F.2d at 764–65).

The fact that the appellate court agreed with the district court on only 8 of the 15 decisions the district court had made in favor of the environmental groups also says something about the uncertainty inherent in the judicial review process and raises most directly the question of what the court is supposed to be doing. In both the water supply replacement case and the fugitive dust case, the issue was one of statutory interpretation, just as it was in the case discussed earlier about the scope of the secretary of the interior's rule-making power. There have been other statutory interpretation decisions involving SMCRA (see Drummond Coal Co. v. Hodel, 610 F. Supp. 1489 [D.D.C. 1985], *aff'd* 796 F.2d 503 [D.C. Cir. 1986] [reclamation fees]; Peabody Coal Co. v. Watt, 553 F. Supp. 1201 [D.D.C. 1982] [prime farmlands]). These cases did not involve interpreting coal mine data or technology.

Statutory interpretation issues contrast with the types of issues about which the appellate court said: "This is precisely the sort of issue on which a court should defer to agency expertise" (839 F.2d at 761). This latter instance involved temporary but long-term storage of top soil, for example soil that is removed in the process of erecting a machine shop or a preparation plant. Both a machine shop and a preparation plant probably will continue for the life of the mine. The secretary intended to allow the mine operator to spread that topsoil and to make interim use of it on nonmined locations (30 C.F.R. section 816.22[c][3]; 817.22[c][3] [1983]). The congressional mandate in SMCRA was that the topsoil be maintained, free of erosion and contamination and in a usable condition (section 515[b][5]; 30 U.S.C. section 1265[b][5]). The National Wildlife Federation argued that there was no evidence that interim spreading and use would not damage the topsoil, the host soil, or both. They argued that, at most, an experimental program should be allowed. The secretary of the interior had relied on technical studies already in existence to frame the regulation. His decision was not arbitrary or capricious; his interpretation of the technical data was rational. The issue was "precisely the sort on which a court should defer to agency expertise." Even though the discussion of "contrary to law" is not concluded, this brings the chapter squarely to a discussion of the third question. Is the regulation arbitrary or capricious?

In a case arising under the APA, the United States Supreme Court explained the arbitrary and capricious review standard.

> Scrutiny of the facts does not end, however, with the determination that the Secretary has acted within the scope of his statutory authority. . . . To make this finding [that the agency action is not arbitrary, capricious, or otherwise outside the law] the court must consider whether the decision was based on the relevant factors and whether there has been a clear error of judgment. . . . Although this inquiry into the facts is to be searching and careful, the ultimate standard of review is a narrow one. The court is not empowered to substitute its judgment for that of the agency. (Citizens to Preserve Overton Park, Inc. v. Volpe, 401 U.S. 402, 416 [1971])

In a case arising under SMCRA, the federal fourth circuit court of appeals observed:

> As a general rule of law, agency action is arbitrary, capricious, or otherwise inconsistent with the law unless the record demonstrates that it is rational, based on relevant considerations, and within the scope of the agency's delegated authority. . . . Furthermore, insofar as an agency decision marks a dramatic change in agency policy, it must be supported by a reasoned analysis explaining the

change. (Virginia Citizens for Better Reclamation, Inc. v. Watt, 14 E.L.R. 20,679, 20,680 [4th Cir. 1984])

Clearly the focus is on rationality. Is there a rational basis in the record for saying that the regulation will accomplish the end result? The concern with judicial review in this area always has been the danger that judges will substitute their judgment for that of the agency. Judges are not supposed to do so; the idea is that the agency has or develops the expertise in the area, and when there are matters of judgment dealing with that expertise, the agency's judgment should control. Thus courts usually defer to the agency's decision that a particular regulation will accomplish the desired result. However, the agency must disclose the evidence and the reasoning behind the decision so that a court can determine whether there is evidence and whether the reasoning makes some kind of sense. The appellate court's discussion of the issue about the thickness of overburden, for example, demonstrates how the secretary's decision was inadequate:

In 1983, the Secretary eliminated the numerical definition, permitting a variance whenever the mine operator demonstrates that spoil is either "insufficient" or "more than sufficient" to restore land to its approximate original contour. . . . The sole support we have found for this revision is the Secretary's cryptic observation that "the mathematical limit . . . has proved to be impractical because of its preciseness." . . . We do not know from this unadorned statement why no adjusted (less precise) or alternate nationwide rule was ordered in place of the one found impractical. Absent fuller statement of the reason for the revision, we cannot intelligently determine whether the Secretary has a "satisfactory explanation" for his action. (839 F.2d at 737)

In other words, the court looks for a rational basis or explanation to sustain the agency's decision. The courts are admonished to take a hard look at the record so that they do not go to the other extreme and become mere rubber stamps for agency action.

When it comes to interpreting the law, as contrasted with interpreting technical data, the issue of deferring to agency interpretation has been cloudier. The courts, after all, and not the agencies, are the experts in interpreting law. Therefore, arguably, the courts should not defer to the agency in those instances. However, the courts clearly are giving more and more weight to agency interpretation of statutes. Most of the seven instances in which the federal appellate court overturned the district court were issues of statutory interpretation, with the appellate court emphasizing the rationality of the secretary's interpretation as it did in the following instance:

> We pretermit the question of whether incremental and/or phased bonding can be accommodated within the terms of . . . the Act. Section 509(c), we hold, reasonably could be read by the Secretary to permit his approval of *alternative bonding methods* that will fulfill the purposes of the Act. While §509(c) is not a model of the drafter's art, and the district court's reading of the provision is a plausible one, we defer to the reasonable interpretation ultimately proffered by the Secretary. (893 F.2d at 725; see also Drummond Coal Co. v. Hodel, 610 F. Supp. 1489, 1496 [D.D.C. 1985], *aff'd* 796 F.2d 503 [D.C. Cir. 1986])

Thus, in a situation where two plausible interpretations of the governing statute exist, the agency's interpretation will be preferred. Perhaps this deference will occur only in the situation noted by the federal appellate court in the District of Columbia: "Deference to the administering agency is particularly appropriate when a complex regulatory statute emerges from a process of difficult legislative gestation" (653 F.2d at 522). SMCRA clearly qualifies as "a complex regulatory statute." To the extent that the courts are now deferring to agency decisions on procedure, on interpretation of the law, and on interpretation of the evidence, unless a decision is clearly contrary to the law or is irrational, judicial review of agency regulations should not interfere that much with agency administration, except for the delay attendant to the judicial review process. This delay can be long, however. The secretary of the interior had promulgated revised regulations in 1983, and it was not until 1988 that the appellate court issued its opinion on the validity of those under challenge, and still today some final regulations remain unsettled.

Approving State Primacy

Under SMCRA, any state that has, or anticipates having, surface coal mining operations on nonfederal lands can "assume exclusive jurisdiction over the regulation of surface coal mining and reclamation operations" (section 503[a]; 30 U.S.C. section 1253[a]). To qualify for exclusive jurisdiction, the state must submit a program that meets the criteria set forth in SMCRA.

SMCRA criteria require that a state program demonstrate that state's capability to implement SMCRA, by providing: (1) a law that is in accord with SMCRA, (2) sanctions for violations that meet the minimum under SMCRA, (3) a regulatory authority that is sufficiently funded and adequately staffed, (4) a permit system that meets SMCRA requirements, (5) a process to designate areas unsuitable for surface coal mining, (6) a permit system that avoids duplication with other federal and state permitting processes, and (7) regulations that are consistent with those issued by the secretary of the interior under SMCRA (section 503[a][1]-[7]; 30 U.S.C. section 1253[a][1]-[7]).

The secretary must determine whether a state program meets the SMCRA criteria before approving the program (section 503[b] and [c]; 30 U.S.C. section 1253[b] and [c]). Basically, the seven criteria listed in SMCRA mean that state statutes must be in accordance with federal requirements, and state regulations must be consistent with federal regulations. The secretary of the interior defined those terms as follows:

> "Consistent with" and "in accordance with" mean:
> (a) With regard to the Act, the State laws and regulations are no less stringent than, meet the minimum requirements of and include all applicable provisions of the Act.
> (b) With regard to the Secretary's regulations, the State laws and regulations are no less effective than the Secretary's regulations in meeting the requirements of the Act. (30 C.F.R. section 730.5)

Under SMCRA, the courts are instructed to uphold the secretary's approval or disapproval of state programs, "unless the court concludes that such action is arbitrary, capricious, or otherwise inconsistent with law" (section 526[a][1]); 30 U.S.C. section 1276[a][1]). Four published court opinions review the secretary's approval or disapproval of programs from four different states. None of these involve a challenge to an entire state program. In Pennsylvania Coal Mining Association v. Watt (562 F. Supp. 741 [M.D. Pa. 1983]), the regulated industry complained about three aspects of the Pennsylvania program that the secretary approved and one aspect the secretary disapproved. The court upheld the secretary's decision on three aspects, but overturned his decision on one aspect. In the other three cases, citizen groups complained about the secretary's approval of state programs. In Virginia Citizens for Better Reclamation, Inc. v. Watt (14 E.L.R. 20,679 [4th Cir. 1984]), the complaint involved eight aspects of the Virginia program. The court upheld the secretary's decision on four aspects but overturned his decision on the other four aspects. In Citizens for Responsible Resource Development v. Watt (579 F. Supp. 431 [M.D. Ala. 1983]), the complaint involved nine aspects of the Alabama program. The court upheld the secretary's decision on six aspects but overturned his decision on three aspects. Finally, in Illinois South Project, Inc. v. Hodel (844 F.2d 1286 [7th Cir. 1988]), the complaint involved six aspects of the Illinois program. The court upheld the secretary's decision on four aspects but overturned his decision on two aspects.

Thus, simply on numbers, the secretary of the interior prevailed on 17 aspects and lost on 10 aspects. In all but two of the 10 instances where the complainants prevailed, the court found specific language in either SMCRA or the federal regulations that had not been met in the state programs. Furthermore, in those instances, the secretary had failed to articulate any basis for rationalizing the difference. Thus, the secretary's approval was arbitrary and capricious. Of the other two instances where complainants prevailed, one involved a mistake by the secretary and the other a change of circumstances after the program provision was drafted.

In the instances where the secretary prevailed, the secretary either had explained rationally the differences between federal language and state language or explained away the perceived difference as not being real. In either event, the process resulted in clarifying the meanings of the approved portions of the state programs.

In effect, in the latter instances the court read the approved program to be consistent with the substance that the complainants were arguing for and concluded that the complainants had not read the state program correctly. Two good illustrations of this come from the Illinois case. First, the complainant was objecting to OSMRE approval of the Illinois regulation that provided: "The highest coal seam may be covered with a minimum of four (4) feet of water" (Ill. Admin. Code section 1816.103[a][1]). The federal regulations clearly required that the minimum depth be 10 meters, a distance considerably more than four feet. However, in the administrative review process, Illinois had assured the secretary of the interior that Illinois read the four-foot requirement to mean "no less than ten meters" (49 *Federal Register* 13,514 [1984]). The court determined that because the secretary interpreted the Illinois regulation in that way, that is what it means; therefore, the complainants had nothing to complain about: "a rule means what it has been interpreted to mean (though it would certainly help things to revise the text)" (844 F.2d at 124).

Another Illinois regulation provided that "box cut spoils shall blend with undisturbed land with a maximum outslope steepness of twenty-five (25) percent" (Ill. Admin. Code section 1816.71[g][2]). Complainants interpreted this state regulation to mean that a reclamation area would blend with undisturbed land as long as it did not contain box cut spoils with maximum outslope steepnesses that exceeded 25 percent. The court said that this interpretation was erroneous and that the regulation established blending as a prerequisite. If no hills exist in the undisturbed area, none can be created in the reclaimed area. If hills do exist, then for box cut spoils a reclaiming party can blend to the undisturbed land by creating hills; but in no event can the slopes exceed 25 percent in steepness.

In effect, complainants prevailed on these two arguments—not by getting the regulations thrown out, but by getting a judicial interpretation consistent with what they believed the regulations should have provided in the first instance. As the court in the Illinois opinion points out, these judicial interpretations are binding, and the agency cannot unilaterally change them:

> All language conceals ambiguities. What the Secretary has approved is not the language in the raw, but the text as interpreted. To give it any other meaning in the future is to amend the rule; Illinois may not do this without a new submission to the Secretary. If Illinois should change its regulation—by promulgating a new one or by reinterpreting the old one—without the Secretary's approval, Illinois South would be entitled to a declaration from the district court that the State was not in compliance with the Act. (844 F.2d at 1291–92)

Obviously the clarification is useful to the extent that it avoids controversy later on.

Overseeing State Administration and Enforcement after Primacy

SMCRA provides that when a state assumes primacy, its jurisdiction in administering SMCRA and the federal regulations promulgated under the act is exclusive, except as provided in sections 521 (30 U.S.C. section 1271) and 523 (30 U.S.C. section 1273; section 523 relates to federal lands and will not be discussed further) of the SMCRA (section 503; 30 U.S.C. section 1253). Section 521 contains many subdivisions.

Under a part of the first subdivision of section 521 (subdivision [a][1]; 30 U.S.C. section 1271[a][1]), when the secretary of the interior learns of a possible lack of enforcement, the secretary is to notify the state and, if the state fails to act or to explain its inaction within 10 days, the secretary is to conduct a federal inspection. Under another part of the first subdivision of section 521, the secretary may intervene directly against the mine operator if, after any authorized federal inspection, the secretary determines that there is an imminent danger to public health or safety or imminent environmental harm to land, air, or water (subdivision [a][2]; 30 U.S.C. section 1271 [a][2]).

Under the second subdivision of section 521 (subdivision [b]; 30 U.S.C. section 1271[b]), the secretary of the interior may assume primacy (partial to full administration) of a state program, if violations exist that result from the state's failure to enforce all or part of its program. Such an assumption requires a formal process, including notice and a public hearing.

Controversy has arisen over whether the secretary of the interior has authority to proceed against a mine operator directly when there is neither imminent danger nor an assumption of primacy either wholly or in part. In Clinchfield Coal Co. v. Hodel (640 F. Supp. 334 [W.D. Va. 1985]) and United States v. Camp Coal Co. (637 F. Supp. 336 [N.D. Ala. 1986]), the courts concluded that no authority exists. On the other hand several courts, as in Annaco, Inc. versus U.S. Department of Interior (675 F. Supp. 1052 [E.D. Ky. 1987]), have concluded that authority does exist.

A problem arises with the conclusion of the first two courts that the secretary lacks authority: While the courts' reading of the SMCRA is a plausible one, the secretary's regulations provide otherwise and have been left intact by the secretary after extensive review (30 C.F.R. section 843.12[a][2]). The secretary's reading also is a plausible one. Furthermore, the time for appealing the validity of the regulation has passed, and circumstances seem not to have changed (section 526[a][1]; 30 U.S.C. section 1276[a][1]; within sixty days of the action unless "based solely on grounds arising after the sixtieth day"). Thus, the only question should be whether the regulation is open to collateral attack. On appeal in the Clinchfield Coal Company case, the district court's decision was reversed (802 F.2d 102 [4th Cir. 1986]). The appellate court determined that the district court did not have jurisdiction to rule on the validity of a regulation. The validity of regulations may be challenged only in the federal courts in the District of Columbia.

All three of the federal district court decisions can be explained satisfactorily without involving the issue of whether the secretary's regulation is correct. Clinchfield Coal Company and Camp Coal Company can be explained on the basis that Virginia and Alabama properly relied on their secretary-approved state programs to conclude that there were no violations. In Clinchfield Coal Company, Virginia determined that a drainage ditch placed at the interface between fill material and natural ground was allowed under the Virginia program, whereas the secretary of the interior argued that the ditch had to be completely off the fill. In Camp Coal Company, Alabama determined that the time for a mine operator to fill rills and gullies had not expired under the Alabama program and that, therefore, no violation existed under the Alabama program. These determinations by the states should constitute good cause for their failure to act. Both the statute and the regulation accept good cause as an excuse for state inaction. On the other hand, in Annaco, Inc. clear violations occurred that everybody recognized. Annaco had failed to reclaim several parcels of land. The subsequent settlement between Kentucky and Annaco simply did not provide for the reclamation of all of the parcels. Arguably, Kentucky failed to carry out a part of its state program. Thus, the secretary of the interior was free to act under SMCRA section 504(b):

> In the event that a State has a State program for surface coal mining, and is not enforcing any part of such program, the Secretary may provide for the Federal enforcement, under the provisions of section 521, of that part of the State program not being enforced by such State. (30 U.S.C. section 1254[b])

That is exactly what the secretary's regulation does—it provides for federal enforcement.

While it might be argued that in order to carry out 504(b) the secretary has to formally assume partial administration of the state program under section 521, discussed earlier, that argument fails: A third part of the first subdivision of section 521 (subdivision [a][3]; 30 U.S.C. section 1271[a][3]) states that the secretary of the interior may proceed directly against an operator in a situation where danger is not imminent, when there has been a federal inspection pursuant to "a Federal program or a Federal lands program, Federal inspection pursuant to section 502, *or section 504(b) or during Federal enforcement of a State program in accordance with subsection (b) of this section [521]*" (emphasis added; 30 U.S.C. section 1271[a][3]). Obviously if section 504(b) could be implemented only through the assumption of primacy part of section 521, the foregoing excerpt showing them to be alternatives would make no sense. There should be no question that the secretary has authority to directly enforce SMCRA in face of a state failure to do so; however, controversy probably will continue until more definitive appellate court determinations are made. Furthermore, it is important for the secretary to have this power of direct enforcement so states will not lapse into the situations that existed before SMCRA was enacted. Congress enacted SMCRA because Congress perceived preexisting state programs and actions generally were inadequate. As the Annaco, Inc. case shows, it would be easy for that pattern to recur through lax state enforcement.

Judicial decisions not only determine the outcome of particular controversies, but also give guidance to the agency and to courts that may review agency actions in the future. A recent development in the SMCRA litigation has been negotiation of settlement agreements between the parties to the litigation.

In 1986, the National Wildlife Federation and others sued the Kentucky regulatory authority, claiming that the state had not properly administered and enforced SMCRA requirements since it had assumed primacy (National Research Council, Surface Coal Mining Effects on Ground Water Recharge 21–22 [1990]). The case against Kentucky was settled, and when OSMRE, which had been sued separately for breach of its oversight role, agreed to that settlement, the suit against OSMRE was dismissed (National Research Council 1990, 21). A specific item included in the settlement agreement dealt with hydrologic balance and demonstrates how innovative approaches can be used in settlement agreements. The SMCRA provision for the restoration of recharge capacity had been particularly difficult for OSMRE and the state regulatory authorities to deal with: "Minimize the disturbances to the prevailing hydrologic balance . . . [by] restoring recharge capacity of the mined area to approximate premining conditions" (section 515[b][10][D]; 30 U.S.C. section 1265[b][10][D]). Under the settlement, a steering committee was to be selected to advise OSMRE on a study of hydrologic balance. Specifically, that study would undertake to "identify . . . cost-effective approaches to determination of the pre-mining recharge capacity for both Eastern and Western Kentucky coal field regimes" (National Research Council 1990, 21). The steering committee contracted with the National Research Council to study the issue of recharge capacity. The result was a book on the subject (National Research Council 1990).

Another more recent settlement came in the case involving the applicant violator system (AVS), described in more detail in the next section. Here, in contrast with the Kentucky settlement where all of the interested parties were in agreement, industry has appealed dismissal of its objections to the settlement (*Coal Week* 1990, 8).

The parties to any settlement base their decision to settle, in part at least, on their understanding of how previous judicial review cases have fared. The role for the court in the instance of a settlement agreement is to approve the settlement before it dismisses the case. This role is important in cases where both the public interest and fairness are at issue. The public interest aspect of coal mining and reclamation is obvious from the existence of SMCRA itself. The fairness issue comes into play from the perspective of a person, group, or entity that is interested directly but may not have any direct say in the settlement because it is not a party to the litigation. At the very least, the court must avoid the collusive lawsuit where the litigation is set up from the beginning between the two parties for the purpose of obtaining court approval of their agreement (see Note 1990).

Although most of the discussion in this chapter has focused on agency action, the Kentucky settlement agreement case shows that agency inaction is subject to judicial review as well. If Congress has given an agency many duties to perform, without enough money or time to carry them all out on schedule, the agency necessarily makes hard choices about what to do first. Litigation can arise once statutory deadlines are missed or some aspect of a strategy appears to be missing. The best illustration of this arises, however, in the next section.

Direct Administration and Enforcement

The secretary of the interior's primary stint with direct enforcement of SMCRA came during the interim period between enactment of SMCRA and the effective dates of state permanent programs (section 502; 30 U.S.C. section 1252). In particular, section 502(e) provided that "within six months after the date of enactment of this Act, the Secretary shall implement a Federal enforcement program which shall remain in effect in each State as surface coal mining operations are required to comply with the provisions of this Act, until the State program has been approved pursuant to this Act" (30 U.S.C. section 1252[e]). Where states fail to develop programs, the secretary must promulgate the equivalent of a program for the state and then undertake to administer that program (section 504; 30 U.S.C. section 1254). A state may decide for reasons of cost, for example, not to develop a program. If a state runs into difficulties in administering its program, the secretary may resume direct enforcement of all or part of that program (section 521; 30 U.S.C. section 1271; the latter is an oversight role and is discussed in the section on oversight of state programs).

Although the secretary of the interior cited numerous alleged violators of SMCRA early in the administration of the act, he did little to penalize those violators. Finally, such a large backlog of cases existed that one environmental group brought suit, alleging that under section 518 (30 U.S.C. section 1268) the secretary has a mandatory duty to assess penalties whenever a cessation order has been issued. In Save Our Cumberland Mountains v. Watt, (550 F. Supp. 979 [D.D.C. 1982]), decided in September 1982, the court agreed. In December, the court issued its order requiring the secretary to collect various civil penalties and eliminate the backlog of pending cases (Save Our Cumberland Mountains, Inc. v. Clark, 18 E.R.C. 1217 [D.D.C. 1982]). Still dissatisfied with the secretary's progress on the whole enforcement front, the judge amended the order in 1985, requiring, among other things, that the secretary establish the "applicant violator system" and make net worth determinations in aid of enforcement (Save Our Cumberland Mountains, Inc. v. Clark, 22 E.R.C. 1217 [D.D.C. 1985]).

The applicant violator system was to be a computerized system, used at least quarterly to determine if any permit applicant controlled, or had controlled, any entity with an unabated cessation order or unpaid penalty under SMCRA. The system was to be fully operational within 24 months of the order. However, problems continued to plague implementation ("Environmentalists to Negotiate" 1989; U.S. General Accounting Office 1989). The direct parties to the litigation reached a settlement, but the district court's dismissal of objections to that settlement and consequent approval of the settlement have been appealed (*Coal Week* 1990, 8).

Under section 510(c) (30 U.S.C. section 1260[c]), if an applicant is in violation of SMCRA or of any air or water pollution laws anywhere in the United States, the permit is to be denied until the applicant shows that the violation has been corrected or, at least, that the process to correct the violation has started. The computerized system that would track SMCRA violations could be used by the states in administering SMCRA under primacy. The case illustrates the unique role the federal government can play in enforcing SMCRA without necessarily treading on state primacy. It would be difficult for individual states to keep track of the multistate operations of a given entity. Only the federal government is in a position to establish a system to deal with the intricacies of coal mine ownership and operation through the multistate, and even multinational, entities that are involved in the coal mining business in the United States.

The net worth determinations were to be made to assist in assessing both penalties and alternative remedies. No less than 25 net worth determinations per month were to be completed, beginning in January 1985.

The court-imposed requirement to establish the computerized applicant violator system and to make net worth calculations both illustrate a high level of intrusion into daily agency affairs that results from judicial review of inaction. It is clear that penalties were not being assessed and collected. The secretary of the interior was, in effect, arguing that there were more important things to do with limited time and budget, and, in the final analysis, making these assessments and collections was economically inefficient. The court's response was that Congress had imposed a mandatory duty. To the extent that there are coal mine operators who simply mine and run without carrying out reclamation, a practical system of detecting such operators is all the more vital.

The case also demonstrates that when an agency misses a deadline and someone sues, the courts may give the agency a new deadline. The agency with limited time and resources will need to expend the time and resources on the deadline set by the court or face contempt. Thus, another consequence of judicial review is that courts set priorities for an agency that has limited time and resources. While some persons may argue that in most of these cases, if not all of them, the agency agrees to the court deadline, the argument is insubstantial. The agency has little choice; if it does not agree to a deadline, a worse deadline may be imposed. Probably no court would listen to the argument that the aspect being litigated is relatively unimportant in the whole scheme and the court should, therefore, at least in the current budgeting picture, impose no deadline. So the agency argues for and accepts the best deadline it thinks it can get.

On the other hand, the secretary of the interior frequently seeks court assistance in the enforcement process. For example, in United States versus Threet (684 F. Supp. 169 [M.D. Tenn. 1987]), the secretary sought an injunction prohibiting the defendant from mining without a permit. The court granted the injunction, even though the mining had ceased. The court found that the defendant had a past pattern of ignoring cessation orders at six mine sites, suggesting the probability of future mining without a permit. With the injunction in place, any future mining without a permit will constitute contempt of court with all of the sanctions that involves. In this context the courts are useful tools employed by the secretary of the interior to further the enforcement process; SMCRA provides that courts are free "to provide such relief as may be appropriate" (section 521[c]; 30 U.S.C. section 1271[c]).

If, in an enforcement proceeding, the alleged violator pursues the regular administrative process and ultimately appeals the result to the courts, the standard for judicial review in SMCRA provides that "the findings of the Secretary if supported by substantial evidence on the record considered as a whole, shall be conclusive" (section 526[b]; 30 U.S.C. section 1276[b]). No significant court opinions under SMCRA exist to discuss in this context. To date, most court enforcement cases have focused on the secretary's authority rather than on whether the defendant was in violation of SMCRA. Doubt now exists whether the arbitrary and capricious standard and the substantial evidence standard are substantively different. (See Motor Vehicle Manufacturers Association v. State Farm Mutual, 463 U.S. 29, 43 [1983]; Sunstein 1983.) Of course the way SMCRA is structured, most enforcement actions should take place in state court; only when states fail to enforce would they take place in federal court.

Standing

In SMCRA, Congress used the same type of language it used in the APA. Section 526(a)(1) of SMCRA (prescribing judicial review for various rulemaking efforts) states: "Any such petition [for judicial review] may be made by any person who participated in the administrative proceedings *and who is aggrieved by the action of the Secretary*" (emphasis added; section 526[a][1]; 30 U.S.C. section 1276[a][1]). With reference to another section of SMCRA, where Congress provided that persons adversely affected may object to an application for a permit (section 513[b]; 30 U.S.C. section 1263[b]), the House committee specifically wrote: "It is the intent of the Committee, that the phrase 'any person having an interest which is or may be adversely affected' shall be construed to be coterminous with the broadest standing requirements enunciated by the U.S. Supreme Court" (2 U.S. Code Cong. and Ad. News 626 [1977]). Thus, Congress itself recognized in SMCRA its inability to rewrite the Supreme Court's interpretation which has made standing a U.S. Constitution Article III requirement and simply made known its intent to allow the broadest standing the courts will permit (Valley Forge College v. Americans United, 454 U.S. 464, 471–76 [1982]). The court generally has not distinguished between the terms adversely affected and aggrieved.

In National Wildlife Federation v. Hodel (839 F.2d 694 [D.C. Cir. 1988]), the court of appeals for the District of Columbia circuit gave a thorough review to the standing issue, considering 70 affidavits of more than sixteen hundred pages. The court earlier had remanded the issue to the district court for additional fact finding and, after that fact finding, devoted 13 pages of the opinion to the issue of standing (839 F.2d 703–716). Of course the variety of substantive regulations under challenge required the extensive review, because a person injured by a failure to reclaim prime farmland may not have an injury related to wildlife and, therefore, no basis for challenging a failure to protect wildlife habitat.

The case did not present any issue as to the standing of industry to challenge the regulations. The economic impact that the regulations have on industry is obvious. This is not to say that an industry association will always have standing. For example, if the complaint is that private property has been taken by a prohibition against mining on steep slopes, there would have to be some showing of ownership of steep slope property that cannot now be mined.

The plaintiff, National Wildlife Federation, relied on the affidavits of members. The attacks on the plaintiff's standing were basically on (1) the members themselves, (2) the identification of the injury, and (3) the speculativeness of the injury. The first item really focused on the constitutional requirement and the third on the necessary nexus between the injury and the agency action, promulgation of (or failure to promulgate) the regulation in question. The second item obviously was relevant to both of the others.

The affidavits were from members such as (1) persons who "live in communities where surface mining operations have occurred," alleging a "fear from ongoing and future mining" (839 F.2d at 707); (2) "residents [who are neighbors of] a nearby mine" (839 F.2d at 712); (3) an owner of prime farmland subject to a coal mining lease (839 F.2d at 713); (4) a renter of prime farmland next to mined land (839 F.2d at 714); (5) an observer of local fish species (839 F.2d at 715); and (6) two wildlife hunters in a mining area (839 F.2d at 715). As to satisfying the constitutional requirement, the court's description of the affidavits relied on by the plaintiff was ample to show that these affidavits were not like the affidavits found to be inadequate in Lujan v. National Wildlife Federation (497 U.S. 871, 110 S. Ct. 3177 [1990]).

In Lujan a five to four majority found the affidavits inadequate because they "state only that one of the respondent's members uses unspecified portions of an immense tract of territory, on some portions of which mining activity has occurred or probably will occur by virtue of the governmental action" (p. 3189; see also Casenote 1990.) In Hodel, as to lack of specificity, the court found numerous faults with defendants' arguments, and no one response stands out. As to speculativeness of the injury and, therefore, its real nexus to the regulation in question, the court looked at the matter from the perspective of plausibility. Was the line of causation argued in the affidavits plausible? If so, that was sufficient. In all instances, on a case-by-case analysis, the court found sufficient specificity and plausibility.

While SMCRA also contains a specific section entitled citizen suits (section 520; 30 U.S.C. section 1270), this provision, too, limits court actions to "any person having an interest *which is or may be adversely affected*" (emphasis added; section 520[a]; 30 U.S.C. section 1270[a]). Otherwise the provision is extremely helpful in demonstrating the intent of Congress as to the permissible scope of citizen involvement.

CONCLUSION

The discussion in this chapter shows the varied nature of federal court cases that review federal agency action under SMCRA. These cases illustrate the following impacts on the administration and implementation of SMCRA: (1) agency actions have been overturned because they did not conform to correct procedure; (2) agency actions have been overturned as arbitrary and capricious; (3) agency actions have been clarified; (4) agency actions have been overturned as not authorized by SMCRA; (5) priorities for action have been established for the agency; (6) long periods of time have elapsed before regulatory approaches have been settled; and (7) agency reasoning and action have been kept open to public scrutiny.

While it is easy to count the types of impacts court review has had, it is much more difficult to assess how critical that review has been to the implementation of SMCRA. The basic role of the courts, given the extreme deference paid to agency actions, is to keep the agencies within the basic limits set in the act; on the other hand, the courts must enforce the agency's mandatory duties. Clearly, the courts are performing these functions with reference to SMCRA. Probably the decision with the most impact was the 1981 court of appeals decision discussed earlier that the secretary of the interior had broad legislative rule-making power under SMCRA.

Some impacts are lasting; some are not. If the court finds a procedural defect, the agency is free to start over and use correct procedure. If the court finds the agency action to be arbitrary and capricious, the agency may be able to remedy the situation. If the agency failed to explain its action adequately, it can now proceed to do so; alternatively it may be able to gather more data in order to substantiate the approach it took. If an agency disagrees with the way the court clarifies a regulation, the agency may be able to revise the regulation. However, if the court decides that the agency does not have authority to do something, the agency's only recourse is to appeal to a higher court or to go to Congress and seek the authority. Similarly, if the court establishes a priority for the agency, the agency must conform or risk contempt.

Furthermore, if there was no judicial review at the time the regulations were promulgated or state primacy plans approved, issues concerning their validity could arise only in the context of individual enforcement actions. It may be a more efficient use of administrative and judicial time to have these issues determined when the agency action occurs. Beyond the constitutional necessity for a day in court, concerned parties also have a better sense of acceptance of the regulations if they have their day in court.

In the meantime, even though the agency prevails ultimately, judicial review causes substantial delay in implementing the particular regulation, and, perhaps, the entire program, depending on how important that regulation is to the overall scheme. This is an inevitable adverse consequence of judicial review, and while it may be possible to streamline the process further, such steps appear unlikely. Within the context of SMCRA, Congress already has limited general review of regulations to the federal courts in the District of Columbia (section 526[a][1]; 30 U.S.C. section 1276[a][1]). However, it has been argued that the absence of the word "only" does not limit other courts from hearing challenges to regulations.

Because agencies know they may have to submit to judicial scrutiny, they tend to build a record accordingly. In all probability, judges reverse enough agency actions to keep the agencies aware of the mission Congress has given them and to keep them to the task of carrying out that mission most of the time. It might appear that those being regulated have substantial self-interest in keeping regulation to a minimum, and thus they would pursue judicial review when they believed an agency was wrong. However, if a regulated party believes the regulatory authority can "make it tough for them," that party may decide not to challenge a seemingly invalid regulation. Therefore, it may be important to have a trade association that pursues issues on behalf of all regulated parties. When, however, a direct enforcement action is being taken against an alleged violator, only that alleged violator can decide whether or not to challenge the enforcement action.

From the perspective of public interest, if the regulatory agency does not have the interest or the capacity to enforce the act, who will? The answer in recent years has been private citizen groups. Congress has recognized this trend by providing for citizen suits. However, the ability of citizens to bring suit resulted solely from the decisions of the United States Supreme Court in Sierra Club v. Morton (405 U.S. 727 [1972]) and United States v. SCRAP (412 U.S. 669 [1973]) to allow groups "standing" to sue on the basis of alleged environmental harm to the group or a member of the group. It is important for both concepts to continue: being able to sue for environmental harm and being able to represent members.

NOTES

I would like to thank James Williams, a third-year law student at Southern Illinois University at Carbondale, for conducting some of the research for this chapter; Bill Hodes, professor, University of Indiana at Indianapolis School of Law, and visting professor, Southern Illinois University at Carbondale School of Law, for his conversations with me about the general scheme of administrative law; and Cy Fox, professor, University of Pittsburgh School of Law, and Pat McGinley, professor, University of West Virginia College of Law, for their review of and comments on the SMCRA discussion in this chapter.

1. Hodel v. Virginia Surface Mining and Reclamation Association, 452 U.S. 264 (1981); Hodel v. Indiana, 452 U.S. 314 (1981). Of course, as the Supreme Court points out in these opinions, it is still possible that SMCRA provisions will

be held unconstitutional *as applied* to the facts of a particular case. See, for example, Blackhawk Mining Co. v. Andrus, 19 E.R.C. 2051 (6th Cir. 1983) where, in an enforcement proceeding, the mine operator challenged the constitutionality of the SMCRA provision that required payment of the assessed penalty into escrow before the operator could challenge the alleged violation. The court upheld the provision as applied.

REFERENCES

Case Comment. 1979. "Judicial Review of Regulations Promulgated Under the Surface Mining Control and Reclamation Act of 1977." *West Virginia Law Review* 81: 785, 786–89.

Casenote. 1990. "Meeting the Requirements of Standing: A Framework for Environmental Interest Groups: Lujan v. National Wildlife Federation, 110 S. Ct. 3177." *Hamline Law Review* 14: 277.

Coal Week. 1990 (November 12), p. 8.

"Environmentalists to Negotiate on Problems with Applicant Violator Systems." 1989 (September 29). Environmental Reporter: Current Developments (BNA) 20: 957.

Gellhorn, Ernest, and Ronald Levin. 1990. *Administrative Law and Process*. 3d ed. St. Paul, Minn.: West Publishing Co.

Irwin, Will A. 1988. "Federal Administrative Review Under the Surface Mining Control and Reclamation Act: Annotated Procedural Guide." *Journal of Mineral Law & Policy* 3: 417.

National Research Council. 1990. *Surface Mining Effects on Ground Water Recharge*.

Note. 1990. "Protecting Environmental Consent Decrees from Third Party Challenges." *Virginia Environmental Law Journal* 10:137.

Schwartz, Bernard. 1984. *Administrative Law*. 2d ed. Boston: Little, Brown.

Sunstein, Cass. 1983. "Deregulation and the Hard-Look Doctrine." *The Supreme Court Review* 177.

U.S. General Accounting Office. 1989. *Surface Mining: Operation of the Applicant Violator System Can Be Improved, GAO/AFMD-89-31*. Washington, D.C.: U.S. Government Printing Office.

8 The Impact of Court Decisions on the Implementation of a State Regulatory Program

Lisa J. Morris

INTRODUCTION

The litigious history of the Surface Mining Control and Reclamation Act of 1977 (SMCRA) (Public Law 95–87, 30 U.S.C. 1201 et seq.) has had a dramatic effect on states that are authorized by section 503 of the act (30 U.S.C. section 1253) to assume primacy (i.e., primary regulation of coal mining operations on nonfederal lands within state boundaries). Rule making by the Department of the Interior's Office of Surface Mining Reclamation and Enforcement (OSMRE) has spawned litigation that now enters its second decade in the United States District Court for the District of Columbia[1] (In re Permanent Surface Mining Regulation Litigation, hereafter referred to as the Flannery decisions).

In order to discuss the Flannery decisions in the context of their impact on state programs, and specifically in the context of the Ohio coal regulatory program, it is important to identify historical litigants and to define the state perspective on protracted federal litigation. Generally, the Department of the Interior has been the defendant; national environmental organizations, such as the National Wildlife Federation, have been the plaintiffs; and national or state industry associations have been intervenors. Major coal companies and some states have also participated as parties to the litigation. Individual primacy states (those that have assumed state regulation of coal mining under SMCRA) typically have been interested nonparties, although primacy states directly regulate and, thus, have the most direct interest in the outcome of litigation that shapes regulatory direction.

Many of the court challenges are uniquely federal. Does the secretary of interior have statutory authority to promulgate a specific regulation? Was the federal Administrative Procedure Act (5 U.S.C. 553 [1970]) complied with, and was there appropriate administrative record to support the federal rule making? Yet, those challenges to federal rule making often directly affect primacy states and change, often drastically, the scope of state regulation.

SMCRA is not unique among environmental legislation based on cooperative federalism, a system in which state and federal governments work together to implement successful regulatory programs. Under SMCRA, states have substantial authority as primary regulators, and the federal government has a defined oversight role to ensure that federal regulatory standards are met. However, opinion differs as to whether cooperative federalism under SMCRA works, and even whether the model of cooperative federalism is flawed (see Bagnall 1986; Squillace 1985; Dernbach 1986). What is certain, however, is that the litigious history of SMCRA frustrates any regulatory partnership and makes efficient implementation difficult.

This chapter addresses a state's perspective on what is, first and foremost, a federal program. States spend the majority of their time reacting to federal litigation; this chapter discusses this reactive process from Ohio's perspective, and it illustrates that the implementation of federal standards through state primacy programs is itself highly regulated and complex.

MODIFICATION OF STATE PROGRAMS

Hard-fought federal legal battles, often over very technical issues such as the appropriate manner in which to dispose of coal mine waste or the engineering requirements for sediment pond design, are but precursors to a cumbersome and time-consuming process of amending state primacy programs. For example, when in 1983 OSMRE revised its initial (1979) rules governing coal mine waste, citizen and environmental groups challenged the agency's modification of a provision related to coal mine waste hauling and conveying. The modification would implicitly have allowed the side or end dumping of coal mine waste, by eliminating an existing requirement for placement of this material in a controlled manner. The court, in In re Permanent Surface Mining Regulation II (round III) (620 F. Supp. 1519 at 1534 to 1538, 22 Env't. Rep. Cas. 2153 [D.D.C. 1985]), found that OSMRE failed to justify the changes, thus rendering the 1983 rule arbitrary and capricious. Citizen and environmental groups also challenged two rules governing coal mine waste disposal in which existing design standards were replaced with performance standards. The court, in In re Permanent Surface Mining Regulation II (round III) (22 Env't. Rep. Cas. 2153 at 2167), found in the case of both rules that SMCRA mandates the establishment of design standards, and replacing those design standards with performance standards violated section 515(f) of the act.

Also in 1983, OSMRE revised its initial (1979) rules governing sediment pond design. Citizen and environmental groups contested OSMRE's use of size designations established by the Mine Safety and Health Administration to adopt less stringent standards for small coal mine waste impoundments. Industry contested OSMRE's rule requiring that all surface drainage from a coal mining operation be passed through a siltation structure. Industry also contested OSMRE's inclusion in the final 1983 rules of specific static and seismic safety factors, as well as foundation and laboratory testing requirements, for all impoundments, regardless of size. Judge Flannery, in In re Permanent Surface Mining Regulation II (round III), remanded the size distinctions among coal mine waste impoundments because OSMRE did not provide adequate technical justification. He remanded the universal requirement to use siltation structures because OSMRE did not adequately address industry's claims of environmental harm in western states. The imposition of specific safety factors for all impoundments was also remanded because interested parties did not receive adequate notice before publication of final rule making.

A successful challenge to a regulation usually results in the regulation being remanded to the secretary of the interior for additional rule making; the secretary usually suspends the federal regulation. A final federal regulation that corresponds to a federal court decision becomes effective only after the required public notice and comment under the Administrative Procedure Act. The state counterpart regulation, untouched by federal litigation, may then be less effective than the new federal regulation. If that situation occurs, primacy states are notified, pursuant to 30 CFR section 732.17(f)(1), to address what has become a state program deficiency.

What may, after years of litigation, advocacy, and final rule making, appear to federal regulators as a finished chapter is a new topic in a primacy state, delivered in the form of a "732 notice" to change the state program. The state has 60 days in which to respond. During those 60 days, the state must thoroughly review its own requirements, confer with field and technical personnel, review the federal rule making, and initially determine whether it concurs that its state program regulation is less effective than the federal regulation. A simple side-by-side comparison of federal and state rules does not afford the required analysis since, among other things, a state's rule is often not structured the same way as the federal regulation, nor is it semantically identical. If, for example, coal operators must design sediment ponds with principal spillways that will safely pass the peak flow resulting from a one-hundred-year, six-hour storm event, a cursory comparison would show whether the standard is the same; but if the requirement is to design a sediment pond with adequate storage capacity, a different, more complex, analysis is required. Such a review can involve numerous legal, technical, and administrative personnel.

Ideally, the 732 notice is contemporaneous with the final federal rule making, although in practice the notice may lag behind by months—or even years. For example, on November 12, 1985, Ohio received notice of deficiencies in more than 55 rules. These deficiencies resulted from federal rule making challenges from 1978 through 1984 (the Flannery decisions) and from the attendant federal rule making. The state of Ohio embarked on massive regulatory changes over the course of the next three years, and final state rule making was completed in late 1988. While the federal government was challenged in litigation on a case-by-case basis and could thus focus its attention on defined issues, the state regulatory agency, by contrast, had to address simultaneously and in a timely manner all of its alleged deficiencies, and finally had to undertake the state rule making process.[2] Meanwhile, the business of regulating coal mining in the state of Ohio continued on a daily basis.

A state submits to the oversight agency proposed regulatory changes to address program deficiencies; then the state is divorced from the review process. The state will not come back into the process until OSMRE notifies it that the amendment needs clarification or is approved or denied. Despite the rather clear processing schedule dictated by 30 CFR section 732.17, the process of federal publication, review, and determination has rarely, if ever, gone as scheduled. The state of Ohio has 10 program amendments pending at the time of this writing, representing program amendments submitted over the course of three years. The state may receive questions months, or even years, after program amendments are submitted; this adds little continuity to the review process. Program amendments pending in the review stage for long periods of time serve only to destabilize a state program in an area that, through the 732 notice, is already identified as deficient. The state, however, can do nothing to address an identified deficiency until the requisite program amendment is approved. Thus, it can be seen that successful challenges to federal regulations are not self-executing and often may take several years to implement on a state level.

Once the state is advised through publication in the *Federal Register* that its program amendments have been approved, the state may begin its promulgation process in compliance with its own particular administrative procedure statute. Each state's rule promulgation process is different and is codified within that state's statutory law. While some states may have cumbersome processes, Ohio has a straightforward and uncomplicated legislative review of state agency rule making. Ohio Revised Code section 119.03 (Baldwin) requires giving public notice of rule making and a public hearing; filing proposed rules with the secretary of state and with the state's Legislative Service Commission; and holding a hearing before the legislature's Joint Committee on Agency Rule Review, which has the authority to recommend to the legislature that proposed rule making be invalidated if it exceeds the scope of its enabling legislation. If rules go unchallenged or unchanged during the process, a final state rule may become effective within 71 days of the initial filing.

What is not accommodated in the federal system of program amendments under 30 CFR section 732 is the fact that rule making (on both the federal and state levels) is a dynamic and public process that often nets a final product distinct from the initial rule filing. Just as federal rule making and the litigation related to SMCRA have their own constituents and advocacy groups, so does state rule making have distinct parties in interest who actively participate, since active public participation is contemplated, and indeed solicited, in the state rule making process.

Rule making in Ohio typically involves three parties—the state regulatory agency, industry groups or individual coal companies, and environmental groups. The participation can expand, however, depending upon the regulatory topic. In 1989, underground mining and, in particular, longwall mining were the focus of a select legislative committee and of the governor of Ohio. Longwall mining is a method of underground mining in which all coal is removed, resulting in immediate subsidence, or caving, of the overlying strata.

Several Ohio citizen groups have been formed since 1985 to allow citizens to participate as commenters in the processing of specific underground coal mining permit applications. Participation is afforded citizens under O.R.C. section 1513.07. The first group formed was Citizens Organized Against Longwalling (C.O.A.L.), followed by Protect Our World (P.O.W.) and Members Against Subsidence Harm (M.A.S.H.). All remain active as citizen advocacy groups. These citizen groups focused public and political attention on the issue of subsidence. They challenged the regulatory agency's perspective that the right of subjacent support is a property right which the state agency has no authority to regulate;[3] the state coal mining statute does not convey to the regulatory authority the right to adjudicate property right disputes. Homeowners in the region of Ohio subject to longwall mining objected to Ohio Administrative Code section 1501:13–12–03, a regulation that allowed coal operators with clear deed waivers to subside residences without any regulatory obligation to repair or compensate (O.A.C. section 1501:13–12–03[D][2][a] and [b], 1989). Citizens pointed out that deed waivers are often decades old and are contained in the original severance documents that date from the time the mineral estate was severed from the surface estate; current residents are often successors in interest and were not parties to the original severance. In this instance, the governor concurred in the objection and issued a governor's proclamation of emergency and an executive order that supported emergency rule making without public notice and comment. In response to the governor's initiative, Ohio promulgated first an emergency rule, and then a final permanent rule, all without OSMRE program amendment approval.[4]

The final permanent subsidence rule deleted any reference to deed waivers and substituted in its place the ability of current residents and coal companies to enter into agreements for repair and/or compensation (O.A.C. section 1501:13–12–03[E]). No parallel provision exists in federal regulation. Clearly, in this instance, the state political process is not acknowledged in the federal scheme, in which the state has no authority to regulate coal mining without the approval of the director of OSMRE. The result is that Ohio has a regulation, with all the force of any approved Ohio regulation, but the director of OSMRE does not yet recognize the rule, since the program amendment has not been approved. Approval of Ohio's program amendment concerning subsidence control rules is, no doubt, complicated by the most recent litigation addressing subsidence control (National Wildlife Federation v. Lujan, III, [round I] 733 F. Supp. 419, 31 Env't. Rep. Cas. [D.D.C. 1990]).

Many instances of state rule making do not so clearly illustrate that the state regulatory process is distinct from the federal process. As mentioned above, public notice and comment are the foundations of both the state and federal rule making systems. Ohio is required to notify the public no more than 50 days and no fewer than 30 days in advance of a public rule making hearing, legislative in nature, at which any interested party may comment on a proposed rule (O.R.C. section 119.03[A][4]). It is Ohio's experience that, particularly in state rule making resulting from Flannery decisions, industry and, sometimes, environmental groups are active commenters. Since similar national parties initially felt strongly enough about the issues to litigate them, the trickle-down effect on state constituent groups is not surprising. Naturally, the state is obligated to consider public comments, and these sometimes result in negotiated changes to what is already an approved state program amendment, given that rule making follows program amendment approval. When this occurs, an almost endless cycle of point-counterpoint begins, since any substantive change must again be submitted to OSMRE through the program amendment process for approval before the state can promulgate an administrative rule in final form. As an illustration, promulgation of Program Amendment number 25 (PA 25), in which Ohio addressed regulatory reform review I issues (see n. 2, this chapter), was initiated in February 1988. Comments received in public hearings on PA 25, held in March 1988, resulted in revisions to 17 of the 55 rules. The revised rules were refiled in April and early May 1988 and formally submitted to OSMRE as PA 34 on May 26, 1988. On May 27, 1988, OSMRE approved PA 31, which revised provisions of nine rules, all of which were already in the midst of promulgation in PA 25 revisions. With one exception, all 55 of the PA 25 rules were submitted for final filing in early June 1988, with a scheduled effective date of October 1, 1988, as suggested by OSMRE. Errors noted in the versions of five rules submitted for final filing were addressed through refilings in June, August, and September 1988 and submitted to OSMRE as PA 36 on October 20, 1988. PA 34 was conditionally approved by OSMRE on December 22, 1988, and PA 36 was approved on July 25, 1990. Thus, it can be seen that managing large numbers of administrative rules in the process of change, all of which are interconnected by cross referencing, becomes an almost impossible task.

PROGRAM IMPLEMENTATION

Following final state rule making, one would think that the process—which may have begun years earlier in federal litigation—would also come to an end. This is often not the case, for just as a federal regulation can be judicially challenged, so too can a state rule; and just as Judge Flannery can rule on the meaning of SMCRA, so too can a state court interpret a state coal mining regulation. One such case involves associated coal facilities in Ohio.

State litigation, either administrative or in the state court system, always has the potential to bring a state program into question and, perhaps, poses some of the most delicate oversight questions for state and federal agencies. A cautioned, reasoned reaction to state litigation is essential to state program stability.

As a result of Judge Flannery's decision in In re Permanent Surface Mining Litigation II (round I) (Env't. Rep. Cas. 1193 [D.D.C. 1984]), coal crushing, screening, and sizing came under the jurisdiction of both federal and state regulators as part of coal mining operations. In Ohio, such activity was previously unregulated since it was not defined as coal mining. As a result of the Flannery decision, Ohio added approximately 38 associated facilities to its inspectable units list by requiring off-site coal preparation plants and support facilities to obtain coal mining permits or to reclaim affected sites in conformance with Judge Flannery's definition of coal mining (52 FR 17730 [May 11, 1987]). Among those units were several coal dock facilities located on the Ohio River. One such facility challenged Ohio's jurisdiction and carried its appeal of the state's regulatory order to obtain a coal mining permit through the administrative process and on to the Lawrence County Court of Appeals. In Ironton Coal Company v. Division of Reclamation (Case No. 1839, Unreported, Lawrence County, 1987) little deference, if any, was given to Judge Flannery's opinion of what constituted a coal mining operation.

The Ironton court overturned the administrative ruling on the operation of this coal dock facility where coal was crushed prior to being loaded onto barges in the Ohio River. The administrative ruling was clear that this dock facility should be regulated as coal mining, since the dock was operated in connection with a coal mine and coal was processed on the site. Either processing coal or conducting an activity in connection with a coal mine would bring the dock facility under the statutory definition of a coal mining operation under the state program (O.R.C. section 1513.01[G]), or under the federal standard as interpreted by Judge Flannery. The court of appeals, however, disregarded any federal litigation on this subject and, not surprisingly, looked only to its interpretation of Ohio statutory language defining a coal mining operation. The court ignored Judge Flannery's interpretation of identical federal language.

Not only did the Lawrence County court find that the Ironton dock was not operated in connection with a coal mine, it also determined that, although the crushing of coal is physical processing, such processing must be in situ or must occur at or near the mine site in order to be regulated. That decision clearly conflicts with Judge Flannery's ruling on what constitutes a coal mining operation and what the phrase "at or near" modifies. A motion to certify the record of Ironton to the Ohio Supreme Court was overruled, leaving intact in Ohio's fourth appellate district a decision in conflict with the federal program.

Ohio successfully brought 38 facilities under its jurisdiction as a result of Judge Flannery's ruling on what constituted physical processing of coal; one adverse case resulted. As a result of the Ironton ruling, OSMRE initially determined that Ohio's program was deficient and Ohio's statute defining a coal mining operation, although identical to the definition of a coal mining operation in section 701(28) of SMCRA (30 U.S.C. section 1291 [28]), should be amended in order to preclude additional case law development similar to Ironton. Ohio disagreed and argued that one adverse case in a lower appellate court did not constitute a program deficiency that warranted a 732 notice. In the end, OSMRE concurred. To return to its legislature on the basis of one adverse court decision, which did not represent published case law in the state of Ohio and which did not precipitate any additional litigation, was considered by Ohio to be unreasonable and risky. New statutory language could bring the risk of new litigation challenging the meaning of that language.

Programmatic changes in the implementation of SMCRA, and resulting changes in state programs, do not always flow directly from federal litigation and resulting federal rule making. Settlement agreements between federal litigants have, in one instance, prompted perhaps the most controversial changes in the history of SMCRA. The initial consensus between the National Wildlife Federation (NWF) and the Department of Interior in Save Our Cumberland Mountains, Inc. et al. versus William P. Clark et al. (Env't. Rep. Cas. 1217 [D.D.C. 1985] mod'g 18 Env't. Rep. Cas. 184) resulted in what is referred to as the Parker-Gash Order. Under this order, the Department of the Interior obligated itself, and implicitly every primacy state, to enforce SMCRA in a certain manner and to ensure that permits were not issued to individual or corporate violators under the act. NWF alleged that OSMRE was not complying with the Parker-Gash Order and had failed, among other things, to comply with the timetable for implementing the Applicant Violator System (AVS).[5] In an apparent attempt to avoid litigation over noncompliance with the original settlement, the federal litigants entered into a revised settlement in January of 1990 which was, in many ways, more far-reaching among the states.

The revised settlement agreement requires, among other items, that states implement the AVS in a certain manner and that they apply federal regulations that define ownership or control in order to block any permit that would be inappropriate to issue under current federal regulations. Certainly, few would disagree that permits should be blocked for violators. What is remarkable, however, is that a federal agency would obligate a nonparty state in such a manner that its state primacy would be jeopardized if it did not comply with the terms of the settlement, regardless of whether those terms were consistent with individual state law or regulation. This issue brings one full circle to the process of identifying state deficiencies and the program amendment process.

On October 3, 1988, OSMRE promulgated new federal regulations that defined what constitutes ownership or control for purposes of issuing or denying permits to violators of SMCRA or of state coal regulatory statutes. As a result, Ohio received a 732 notice on May 11, 1989. In response, on October 2, 1989, the state submitted an informal program amendment in order to benefit from federal review on a complex program amendment prior to formal submission. OSMRE responded on March 1, 1990, to the informal submission. Despite the attempt to revise its rules, Ohio, like other primacy states, may be months or, more probably, years away from state implementation of the current federal ownership and control regulation, which in the meantime may be dramatically affected by federal litigation that challenges the rule making. This again demonstrates a seemingly endless cycle of point-counterpoint and involves changes in the federal program that require the states to respond.

Another variable in the process of implementing programmatic changes initiated by federal litigation is that of the federal agency's inability, for whatever reason, to promulgate revised federal regulations. A striking case in point is the dilemma both OSMRE and the states face under the rule-making process that surrounds the definition of valid existing rights (VER). The definition of valid existing rights is an essential ingredient in the application of what are referred to as the 522(e) prohibitions under SMCRA (30 U.S.C. section 1272[c]), which prohibit or restrict mining operations in certain areas. Because the manner in which a regulatory agency defines valid existing rights can make a considerable difference in determining which operators are permitted to mine in otherwise prohibited areas, the definition naturally has been hotly debated and litigated.

In looking at its history, federal rule making on VER can be viewed as a total failure; in 1990, some twelve years after the initial rule making to implement SMCRA, no federal definition of valid existing rights exists, and debates continue on how the 522(e) prohibitions should be implemented.

The term valid existing rights was first defined in March 1979. It included those property rights in existence on the effective date of SMCRA, August 3, 1977, where the owner had obtained, on or before the date, all federal and state permits necessary to conduct coal mining operations. This style of definition was referred to as an "all permits test" for VER, since the overriding standard was possession of all necessary permits.

Industry challenged the definition as unnecessarily restrictive, and in In re Permanent Surface Mining Regulation Litigation I (round I) (No. 79–1144, Mem. Op. at 20 [D.D.C. February 26, 1980] 14 Env't. Rep. Cas. 1083) the requirement that the mineral owner obtain all permits was remanded. The court ruled that a good-faith attempt to obtain all necessary permits before August 3, 1977, should suffice. In August 1980, OSMRE responded to the remand by suspending its definition insofar as it required all permits to be obtained, and it stated its intention to interpret the all permits test as including good faith efforts to secure permits.

In June 1982, OSMRE proposed three options for a modified all permits test. Each of the three options provided for "continually created VER," that is, a finding of VER for a coal mining operation which, regardless of its date of origin, existed or was permitted before a 522(e) prohibition came into effect.

In final rule making published in the *Federal Register* in September 1983, OSMRE discarded the three proposed "modified all permits" options in favor of a "takings" test. VER would be found if applying any 522(e) prohibitions to property rights in existence on August 3, 1977, would effect a taking that would entitle the person intending to mine coal to compensation, pursuant to the fifth and fourteenth amendments to the U.S. Constitution. The final rule also retained from previous rules a finding of VER if coal is needed for and adjacent to an operation in existence on August 3, 1977. The concept of continually created VER was also retained, although it was expanded to apply to situations where property rights, and not necessarily a coal mining operation or approved permit, existed before the 522(e) prohibition came into effect under SMCRA.

Citizen and environmental groups challenged the 1983 final rule, arguing that the definition of VER was promulgated without adequate notice and comment under the Administrative Procedure Act, and that the concept of continually created VER is inconsistent with SMCRA. In In re Permanent Surface Mining Regulation Litigation II (round III) (No. 79–1144 Mem. Op. [D.D.C. Mar. 22, 1985], 22 Env't. Rep. Cas. 1557), the court remanded the rule for inadequate notice. The court upheld the concept of continually created VER for property rights, but it remanded that portion of the definition as well, because it related to the takings test and the public notice defect.

In November 1986, OSMRE responded to the remand by suspending its 1983 final rule. Because the suspension would leave federal program states with no criteria for determining VER, however, OSMRE stated its intention to resurrect the original 1979 rule and apply it as modified by the 1980 court decision to include good faith efforts to obtain all necessary permits. For federal lands within states that have approved state programs, where OSMRE retains the responsibility for VER determinations, OSMRE intended to comply with the state definition if it contained the all permits test, but if the state definition required a takings test, OSMRE would not process VER applications until a final federal regulation was promulgated.

In its appeal, the National Wildlife Federation contested the court's March 1985 decision that the concept of continually created VER is in accord with SMCRA (National Wildlife Federation et al. v. Hodel, et al., 839 F.2d 694 [D.C. Cir. 1988]). While the National Wildlife Federation argued that continually created VER is an oxymoron and contrary to legislative intent, the mining industry argued that a state regulatory agency's authority to infringe on mining rights because of a later declaration of unsuitability according to 522(e) would effect a taking. The appellate court affirmed the lower court's decision, thus upholding OSMRE's regulation as reasonable.

The most recent round of federal rule making on VER was initiated in December 1988. However, OSMRE withdrew its proposed rule on July 21, 1989, and has not yet initiated new rule making on this topic. Because OSMRE has been unable to promulgate a revised definition of VER does not mean that states have not been able to define VER and implement the corresponding state counterpart of the 522(e) prohibitions. Depending upon the state definition and whether it would implicitly be affected by a 1985 court decision or explicitly targeted by a 732 notice, that definition may remain unscathed by the federal turmoil. Such is the case in Ohio, where there were no defects in public notice and no components of a takings test in its definition—two problems with the federal definition. In fact, Ohio has been relatively stable and consistent in implementing mining prohibitions, but that stability is threatened by the continuing debate and instability on the federal level.

In a system that is interconnected, yet separate, as is the state regulatory program—a program that operates pursuant to state law, but only with the approval of the secretary of the interior and subject to constant oversight scrutiny—it is naive to expect that a regulatory issue that appears settled within state boundaries will remain settled if the issue is unresolved on the federal level. Complicating the VER issue is the fact that federal lands exist within Ohio's boundaries, and the state regulates those lands pursuant to a cooperative agreement between the secretary of the interior and the governor of Ohio. Under section 523(c) of SMCRA (30 U.S.C. section 1273[c]), the determination of VER on federal lands is a nondelegable responsibility of the secretary of the interior. Thus, the stage is set for inconsistency within state boundaries. In fact, a permit to mine federal lands within Ohio's boundaries was issued, based on the secretary's interpretation and application of the state's definition of VER; if the state had made the determination, the result would have been different. Under Ohio's definition of valid existing rights, the state does not recognize the transferability of VER; that is, the rights claimed must be the applicant's. OSMRE, on the other hand, recognizes transferability. Thus, Ohio issued a coal mining permit (permit D–557–1) to Avis Coal, to mine portions of the Wayne National Forest. The permit was issued on the basis of OSMRE's determination that Avis's lessor, Sunday Creek Coal Company, possessed VER. If Avis had claimed VER through Sunday Creek Coal Company in order to mine nonfederal land in Ohio, the Ohio Division of Reclamation would have denied VER to Avis (see O.A.C. section 1501:13–1–02[FFFFFF]). The subject of VER is, at best, fraught with uncertainty, and states will be unable to resolve this uncertainty until the federal regulation defining VER is finalized.

PROGRAM CONTINUITY AND PRIORITIES

Coal mining regulation is a complex system of federal-state interplay which is not confined to comparing rules and regulations, litigation, and political climates. Numerous variables—from the relationship of the state's legislative and executive branches, to the role of citizen or environmental groups that are often concerned only with specific issues—affect the climate in which a state regulatory agency must attempt to build a stable regulatory environment. Certainly the challenge to implement SMCRA consistently throughout the states deserves a working partnership between the federal oversight agency and the state regulatory authorities. The general consensus among state regulatory authorities, however, is that the partnership is dysfunctional (Harris and Close 1982). Squillace (1985) suggests that OSMRE should turn the oversight table—that the federal government should assume direct regulation and the state should fill the oversight role.

Federal litigation under SMCRA challenges both state and federal governments and produces both actual and perceived instability. The actual instability flows from federal court remand and suspension of federal regulations, and the perception of instability flows from the uncertainties surrounding long-term federal litigation. Program states are required to implement federal regulatory changes through the state program amendment and rule making process even when the federal regulation has been challenged in federal court. States know that such challenges often take years to resolve and that, once resolved, there is a good chance the litigation will result in a suspended or remanded federal rule. In the meantime, states must promulgate state counterpart rules and define such rules knowing that a federal judge may ultimately disagree. This process seems to be never-ending when the state accepts the passive role thrust upon it while a federal judge changes the state's regulatory course. Ohio, in an effort to strengthen a working partnership with the federal government, has determined that the state should aggressively pursue intervention in federal lawsuits where appropriate and in other cases file *amicus curiae* briefs. By taking a more active position in federal litigation, Ohio also would hope to become proactive, rather than reactive, as a regulator.

CONCLUSION

State coal mining regulatory programs are extremely difficult to administer. They span variations in physical climates, from Montana to Alabama; variations in mine size and coal production; and variations in size of regulatory staff. A number of states had some coal mining regulation that preceded SMCRA, and a few, like Ohio, may have regulated coal mining for decades.[6] SMCRA gave the states the opportunity to build and strengthen existing programs. Regardless of a state's experience before SMCRA, it would be difficult, if not impossible, to find any state regulatory agency that questions the necessity of a federal environmental act that requires reclamation of currently mined land and that funds the reclamation of abandoned mine lands.

Nonetheless, state regulators live in a state of constant frustration since a state is often off-center and reacts to such strong influences as federal litigation, federal settlement decisions, and federal rule making. Certainly, Congress intended for OSMRE to set under SMCRA the federal standard for regulating the coal mining industry, but where numerous state and federal interest groups, differing political agendas, and regulatory agencies come together, it is difficult to agree on that federal standard. As a result, in the state forum, often there is no consistency and little opportunity to manage priorities, and survival of the regulating agency may take precedence over protection of the environment.

Even though the OSMRE is responsible under SMCRA for setting the national standards for coal mining and reclamation, until states and the federal government can come to terms with the idea that they must truly function as a partnership, coal mining regulation cannot be successfully implemented. Within that partnership, all parties should recognize that we are all ultimately interested in the same litigation, the same rule making and, although we may take different paths, the same results.

In the meantime, the federal litigation continues, and the states wait in the wings.

NOTES

1. Venue for judicial review of federal rule making is governed by section 526 of SMCRA, 30 U.S.C. section 1276.

2. The 1985 program notice involving 55 rules is referred to as "regulatory reform review I." OSMRE, since 1985, has served the state of Ohio with four additional 732 notices, two of which, referred to as "regulatory reform II" and "regulatory reform III," addressed final federal rule making over successive periods of five years and one year, respectively. Each regulatory reform package in the state forum involves a parallel state regulatory process which is not technically a component of the federal rule-making process and is not recognized under the terms of SMCRA or federal program amendment regulations that implement SMCRA.

3. The right of subjacent support is generally described as a property right attendant to the surface estate such that the owner of a mineral estate cannot subside the surface without liability to the surface owner. Subjacent support is defined as "the right of land to be supported by the land which lies under it" (*Black's Law Dictionary*, West, 5th ed., 1979). The right of subjacent support is a property right which can be conveyed to the mineral owner, who would then possess the right to subside the surface.

4. A summary of the history of Ohio's program amendment on subsidence control is as follows:

1/18/89: As a result of Executive Order 89–4, emergency rule 1501:13–12–03, in which the deed waiver provision was stricken, became effective.

2/1/89: A version of rule 1501:13–12–03, identical to the emergency rule, was submitted for original filing to make permanent the deletion of the deed waiver provision.

2/2/89: Ohio Program Amendment number 38, a version of rule 1501:13–12–03 identical to the emergency and original filings, was submitted to OSMRE.

4/18/89: Emergency rule 1501:13–12–03 expired.

5/5/89: As a result of Executive Order 89–31, a second emergency filing of rule 1501:13–12–03, identical to the first, became effective.

7/7/89: The original filing of 1501:13–12–03 was withdrawn and refiled to substitute a revised version that included language regarding private agreements between landowners and coal operators.

8/8/89: The second emergency rule 1501:13–12–03 expired.

8/9/89: The version of rule 1501:13–12–03 refiled on July 7, 1989, was submitted for final filing.

8/11/89: Revised Program Amendment number 38, a version of rule 1501:13–12–03 identical to the refiled version, was submitted to OSMRE.

8/19/89: Permanent rule 1501:13–12–03, in which deed waiver language was deleted and private agreement language was added, became effective.

5. The applicant violator system is a computerized system that links primacy states and the OSMRE; it is designed in concept to implement the requirements of 30 U.S.C. section 1260(c), federal regulations promulgated thereunder, and the requirements of approved state programs.

6. Ohio first regulated coal mining in 1948 under legislation that required bond at a rate of one hundred dollars per acre and a coal mining permit that described the location of the mine and the names of the landowners. Reclamation under the 1948 law consisted of covering the exposed coal seam; grading to level peaks and ridges; and planting trees, grasses, or shrubs. Ohio's coal mining law underwent only minor changes until it was significantly revised in 1972 to adopt meaningful permitting information requirements, coal mining performance standards that would require protection of the environment while mining, and reclamation standards that would ensure productivity of the land after mining. For the first time, operators were obligated to backfill the entire highwall. In 1981, Ohio responded to the enactment in 1977 of SMCRA, which was based in part on Ohio's 1972 law, by amending some provisions of its law. Perhaps the most significant of the 1981 amendments provided for regulating the surface effects of underground mining.

REFERENCES

Bagnall, Kerry. 1986. "Regulatory Federalism and Its Impact on State Sovereignty." *Capital University Law Review* 15: 561.

Dernbach, John C. 1986. "Pennsylvania's Implementation of the Surface Mining Control and Reclamation Act: An Assessment of How 'Cooperative Federalism' Can Make State Regulatory Programs More Effective." *University of Michigan Journal of Law Reform* 19: 903.

Harris, James R., and Carl C. Close. 1982. "Redefining the State Regulatory Role." *Environmental Law* 12: 921.

Squillace, Mark. 1985. "Cooperative Federalism Under the Surface Mining Control and Reclamation Act: Is This Any Way to Run a Government?" *West Virginia Law Review* 87: 687.

IV State-Level Implementation

9 Implementation of the Surface Mining Control and Reclamation Act in Pennsylvania: A Decade Later

John C. Dernbach

INTRODUCTION

It has been said that the Surface Mining Control and Reclamation Act (SMCRA) is based on a paradox; the same states whose overall poor regulatory performance inspired federal legislation were then given primary responsibility for its implementation (Rochow 1978). For Pennsylvania, the paradox was compounded by irony. Although Pennsylvania's program in many ways served as a model for SMCRA and was praised as the best program in the country (Udall 1977), SMCRA's advocates clearly understood that the legislation would force significant changes in the program (Hays 1987).

After Pennsylvania achieved primacy in 1982 to run its coal mining regulatory program under SMCRA, the effectiveness of its program improved dramatically. Program content did not change significantly; concurrent backfilling (filling pits at the same pace they are being opened) and treatment of acid mine drainage, for example, are longstanding features of the Pennsylvania program. But by the mid 1980s, this writer concluded, the program was enforced and administered in a more vigorous and systematic way than it ever had been (Dernbach 1986). As a result, backfilling occurred more quickly, erosion and sedimentation from surface coal mines decreased significantly, the number and severity of new acid mine drainage outbreaks declined, and water supplies received better protection from surface mining. A recent Office of Surface Mining (OSM) oversight report on the Pennsylvania program notes that the program "continues to enhance environmental protection associated with coal mining and reclamation activities" (Office of Surface Mining 1990).

As Pennsylvania's program approaches the end of its first decade under SMCRA, the stronger enforcement and administration of mining laws that came with SMCRA continue to protect the environment. Two parallel developments, however, are shifting much of Pennsylvania's debate over coal mining in a different direction than SMCRA's supporters might have imagined.

First, the most significant issues now are outside the original model of the problem. The national outcry over surface mining in the 1970s was crystallized by photographs of scarred countryside from unreclaimed surface mines. In retrospect, it is easier to fill holes than it is to solve other problems. Almost 15 years later, the most important of these in the Pennsylvania program are the adequacy of bonding, especially for acid mine drainage, and the environmental effects of underground coal mining, including subsidence, acid mine drainage, and loss of water supplies.

Second, the implementation strategy for these issues under SMCRA is contrary to SMCRA's overall approach to implementation. The congressional debate was premised on the view that detailed statutory performance standards, mandatory enforcement, and federal and citizen oversight were necessary to limit the discretion—and ensure the effectiveness—of state programs. The bonding and underground mining issues, however, involve statutory language that is vague, difficult to interpret, or missing. As a result, both Pennsylvania and the federal oversight agency, OSM, must use discretion and judgment to develop solutions.

These emerging issues are difficult in themselves and are likely to shape much of the national coal mining debate before the end of the century. But they also challenge the basic premises of SMCRA.

CLARITY, VAGUENESS, AND PROGRESSIVE IMPLEMENTATION

Government programs can address implementation in two ways—either through programmed implementation (usually by controlling the discretion of the field-level people who actually implement the program) or through adaptive implementation (allowing field-level people to adapt general program goals to what they learn and experience while they implement the program) (Berman 1980).

Advocates of adaptive implementation claim that it allows field-level people to tailor their response in specific cases and facilitates problem solving. Over time, appropriate policies can be developed incrementally in the field (Lindblom 1968; Kirst and Jung 1982).

Conversely, proponents of programmed implementation believe that field-level people use discretion as a mask for inaction and that government officials use it to justify standardless and often arbitrary behavior (Davis 1971). They claim that discretion creates uncertainty about what is required and adds time to government decision making. Some cynics even see the adaptive approach as a way for Congress to separate itself from politically painful decisions or from an agency's eventual failure to solve a problem (Costain 1979).

Programmed and adaptive implementation represent opposite ends of a spectrum. In practice, most statutes contain elements of each: The most programmed statutes, for example, provide room for discretion, and the most adaptive statutes contain limits on discretion.

SMCRA is a programmed implementation statute in relatively pure form. It sets out detailed performance standards and was intended to allow state-level managers relatively little discretion. It goes so far as to establish a mandatory enforcement program that is unparalleled in other federal environmental laws (Dernbach 1986). There are, however, issues that Congress either did not fully understand or did not wish to answer clearly. Some provisions of the law are vague, or even missing—the result, perhaps, of a lack of complete information and a need to develop more concrete policy through adaptive implementation. Because opponents hold widely diverging views, it has taken time to develop consensus. This problem has been exacerbated because the implementing federal agency has failed to provide needed guidance and has flip-flopped on major questions. These represent the primary unresolved issues in implementing SMCRA.

In Pennsylvania, both the state's Department of Environmental Resources (DER) and the Office of Surface Mining Reclamation and Enforcement recognize that SMCRA's numerous requirements cannot all be implemented at once. Implementation has been based on both administrative priorities and the internal logic of the statute. OSM gave priority to DER's enforcement program in the first years after primacy. However, unless they are forcefully pushed to center stage, statutory provisions that require adaptive implementation will not be implemented as rapidly as provisions that require programmed implementation.

Fifteen years after SMCRA was adopted, Pennsylvania's implementation of SMCRA is becoming dominated by issues that involve substantial discretion. This challenges the law's initial premise and has significant consequences for understanding state implementation of SMCRA during the 1990s.

PENNSYLVANIA'S IMPLEMENTATION OF THE ORIGINAL DESIGN

When Pennsylvania adapted SMCRA into its coal mining regulatory program, it blended federal requirements with the best features of Pennsylvania's existing program (Dernbach 1986). As a result, the program's effectiveness has improved in several areas.

Mandatory enforcement: Under SMCRA, inspectors are to conduct 12 inspections of each operation per year and are required to identify all violations in writing. In general, violations must be corrected within 90 days. Pennsylvania has made steady progress in meeting mandatory inspection requirements. OSM's report for 1989-90 notes that 98 percent of the mine sites were inspected at the required frequency, and 97 percent of the violations were corrected within the initially prescribed abatement period (OSM 1990). OSM has concluded that "frequent inspections result in fewer violations and enhanced compliance" and that part of the reason for the improvement is better communication between DER inspectors and mine operators concerning site conditions and program requirements (OSM 1990).

Pennsylvania requires surface mining operators to obtain both an annual license and site-specific permits, and DER cannot issue a permit or license to an operator in violation. DER has developed a centralized computer data base with current information about program violations, and it is considered to be one of the agency's most effective enforcement tools.

Outside scrutiny: The primacy program differs from the old Pennsylvania program because of OSM's direct oversight and because of extensive public participation opportunities. This kind of scrutiny encourages a more balanced program and one that corresponds more closely to the written program.

More professional management: Under SMCRA, DER staff have increased threefold since 1978 because of the OSM grant that accompanied primacy. The agency now has sufficient personnel to conduct the necessary inspections and to give permit applications the kind of careful technical review they deserve.

Written program: While Pennsylvania's old program existed largely as bureaucratic custom and folklore, which made both management and enforcement difficult, the primacy program is written in statutes, detailed regulations, program guidance, and other sources. DER can administer the program in a professional and consistent manner, and the coal industry and the public can understand what is required.

Enforceable rules: Performance standards in the program are based more on mechanical rules than on judgment rules. Such rules are easier to enforce and, therefore, encourage compliance.

Better information: Finally, DER is making better decisions under the primacy program because it has better information. The development over the past decade of improved permit application review techniques has unquestionably reduced the number and severity of problems. The use of customized computer programs and computer data bases greatly assists the state in permitting and enforcement activities. This kind of sophistication has encouraged the coal industry to be more careful in selecting mine sites and preparing permit applications.

PROBLEMS WITH THE ORIGINAL DESIGN

Mandatory enforcement, outside scrutiny, more professional management, a written program, enforceable rules, and better information can all enhance the effectiveness of a properly designed program. These measures can also highlight underlying weaknesses in program design because implementation alone cannot be blamed when things go wrong. Bonding and acid mine drainage, as well as the environmental effects of underground mining, have been identified as problems with the national program (McElfish and Beier 1990). Pennsylvania's experience shows both the importance of these problems and the difficulties of resolving them.

Bonding and Acid Mine Drainage

Both SMCRA and Pennsylvania law require operators to post a bond in an amount sufficient to provide for the reclamation of a site if the operator fails to do so. Pennsylvania continues generally to require a minimum three thousand dollars per acre bond for areas to be mined, with a higher bond for deeper pits. SMCRA also authorizes states to devise alternatives to a per-acre bond that achieves the purpose of a bond. By law, the bond must cover the state's reclamation costs if the operator is unwilling or unable to do so. If the bond is forfeited, it must be used for the reclamation of the site for which it was posted. In addition, posting a significant bond decreases the likelihood that an operator will abandon a site. The alternative bonding system in Pennsylvania is a permit application fee of fifty dollars per acre for surface mines, which is deposited in a special fund and can be used to supplement bond forfeiture proceeds at any site.

The Pennsylvania bonding system continues to be plagued by two substantial problems. It does not generate enough money to pay for the costs of reclaiming land on which forfeitures have occurred, and it provides no money for direct treatment of acid mine drainage at forfeited sites (Office of Surface Mining 1991). The bonding issue is national in scope, but has been substantially complicated in Pennsylvania because of acid mine drainage (McElfish and Beier 1990).

Land reclamation: Pennsylvania's bonding program was the target of two legal actions filed in 1981. In one, landowners of unreclaimed bond forfeiture sites claimed, among other things, that the state's procedure for selecting sites for reclamation was haphazard and unwritten, making it difficult to get their own property reclaimed. In response, DER began reclaiming sites on a "worst first" system, prioritizing sites with the greatest risks to public safety and the environment and selecting contractors through public bidding. In 1987, the state adopted regulations that modified this system somewhat. DER is to reclaim "primacy sites" first, regardless of the risks they present to public safety and the environment. Primacy sites are those for which a new permit was issued after March 31, 1983, the official end of the repermitting period for sites that were operating on July 31, 1982, when the state obtained primacy. Generally, reclamation is done through public bidding. DER is authorized to provide for reclamation of "preprimacy" bond forfeiture sites on a priority basis in one of four ways: (1) full-cost reclamation through public bidding, (2) contract with landowner or mine operator, (3) issuance of a mining permit that includes reclamation of contiguous property, or (4) contract with the landowner under cooperative agreement among DER and the county soil conservation district. Because the last three reclamation methods do not require public bidding, payment ceilings based on the current bond rate or the amount of the forfeited bond are included to avoid corruption. These new reclamation methods for preprimacy sites have been used only modestly, but they create a more flexible and responsive system. Finally, beginning in the 1986–87 fiscal year, the Pennsylvania legislature authorized $2 million per year from the general fund to be spent for the reclamation of preprimacy sites, a task that was projected to take at least 20 years at that funding level.

The more important action, Pennsylvania Federation of Sportsmen's Clubs versus Department of Environmental Resources, was a challenge to the financial adequacy of the bonding system. In 1988, Pennsylvania's Commonwealth Court approved a settlement stipulation and consent decree, which has framed the issues for subsequent discussions. This decree acknowledges DER's decision to divide bond forfeiture sites into primacy sites and preprimacy sites. Funds from the alternative bonding system (the $50 per acre application fee) may not be used to reclaim preprimacy sites. For preprimacy sites, the environmental groups that filed the action acquiesced to the $2 million per year legislative appropriation as long as public funding continues to be provided. If not, DER is obligated to evaluate other funding methods. For primacy sites, DER must maintain a detailed computer data base of bond forfeiture sites; perform detailed evaluations of reclamation costs at each site, including treatment of acid mine drainage; and evaluate annually, for at least five years, the adequacy of supplemental funds from the $50 per acre fee. The decree specifically recognizes the issue of bonding for acid mine drainage and obliges DER to propose solutions.

Reclamation of primacy sites continues to fall behind the rate at which bonds are forfeited at mining sites. This is true even though the primacy forfeitures do not involve off-permit mining, nonconcurrent backfilling, or other features that contributed to the costs of reclaiming preprimacy sites. The history of this issue includes repeated claims by coal operators that DER's land reclamation costs reflect mining practices that no longer occur, are increased by the public bidding process, or are based on the surge of forfeitures that followed primacy. Yet Pennsylvania's first annual report after the Sportsmen's Clubs consent decree noted declining revenues and increased reclamation costs, and it indicated that the alternative bonding fund is not capable of complete reclamation of sites for which bonds are expected to be forfeited (Office of Surface Mining 1990).

As of August 1991, DER's alternative bond fund had $4.8 to $8.7 million in outstanding liability for the reclamation of 1,866 acres (Office of Surface Mining 1991). DER believes that most of this deficit is due to the high cost of reclaiming coal refuse reprocessing operations rather than the cost of reclaiming surface mines. DER has nonetheless identified several means of increasing available funds, such as increasing the $50 per acre application fee to $100 or increasing the bond rate.

From a program management standpoint, the greatest problem is the financial integrity of the alternative bonding system. SMCRA provides little concrete guidance on bond adequacy, partly because little thought was given to the solvency of alternative bonding systems and partly because Congress was concerned primarily with land reclamation. Alternative bonding systems, which usually involve a fee paid into a statewide fund, are attractive to coal operators because they are less expensive than posting a bond. At the same time, their solvency depends on a variety of factors that are difficult to predict accurately, including the forfeiture rate, the condition of sites upon forfeiture, and overall reclamation costs. It is difficult to charge a sufficiently high fee in the face of such uncertainties. The very brief and general language in the statute and the absence of meaningful OSM oversight on this question until recently have meant that the states had relatively little guidance on how to implement such alternative bonding systems. As a result, most of these systems are not generating sufficient funds for reclamation (Beier and McElfish 1989).

In late 1990, OSM issued guidance for evaluating the solvency of alternative bonding systems. Designing a program in Pennsylvania that meets these requirements will be politically and technically difficult, even for land reclamation. Because of the fiscal crisis created by the national recession, moreover, the annual $2 million appropriation for reclamation was not included in the budget for fiscal year 1991–92.

Acid mine drainage: SMCRA requires the bond amount to be sufficient to assure completion of the reclamation plan, and it requires the reclamation plan, among other things, to assure compliance with applicable water quality laws (secs. 508[a][2], [9], and [13] and 509[a]; 30 U.S.C. secs. 1258[a][2] and [13] and 1259[a]). Because DER cannot issue a mining permit unless the operator demonstrates that the proposed operation will not cause acid mine drainage, the cost of treating acid mine drainage is not factored into the bond amount calculation. After forfeiture, DER's reclamation cost estimates have not included treatment of acid mine drainage. The costs of treating acid mine drainage are nonetheless becoming central to the bonding debate in Pennsylvania.

In 1989, DER evaluated acid mine drainage discharges on sites where bonds are being held or where bonds were forfeited (Pennsylvania Department of Environmental Resources 1989). DER is legally prohibited from releasing bonds on sites where effluent limitations for discharges are being exceeded and DER was holding bonds worth some $60 million. Of the 742 discharges studied, only 6 percent met DER's effluent limitations for pH, iron, manganese, and alkalinity. The study found that it would cost as much as $27 million per year to treat these discharges to meet effluent limitations.

To solve this problem, DER has discussed a number of proposals with the Mining and Reclamation Advisory Board (created by the legislature as a sounding board for such proposals). One proposal is to create site-specific trust funds to pay for treatment of discharges for fifty years. Another proposal is to release bonds at sites with "low impact" discharges that violate effluent limitations but have no in-stream impacts, if the operators install passive treatment systems (artificial wetlands to capture metals, for example) and place enough money in a trust fund to treat the discharge for 50 years.

For the coal industry, DER's failure to release bonds because of these discharges is financially painful because bonds are increasingly difficult to obtain and because bonds represent capital that cannot be used in other ways. In late 1990, Pennsylvania's legislature passed a bill (H.B. 329) that, in effect, would have relieved a number of coal operators of their legal responsibility for many of these discharges. Pennsylvania's OSM office and the U.S. Environmental Protection Agency's Region 3 expressed reservations about the lawfulness of the bill under the federal Clean Water Act, stating that discharges must meet technology-based effluent limitations at the source and in-stream water quality standards. The bill would also have changed Pennsylvania law, under which an operator is liable for discharges that it has affected, even if they were caused by previous mining. By requiring DER to prove that an operator actually caused a particular discharge, the legislation would have freed many operators of liability for postmining discharges because DER lacks adequate records on the origin of most individual discharges. The governor vetoed the legislation but stated his intention to work with the coal industry to resolve its problems in more moderate ways.

This issue is likely to be complicated in the future by two other matters. First, it is increasingly clear that there are limits to technology's ability to predict post mining discharges; problems still occur. Failure to predict discharges as part of the permit application process results in the creation of more sites that will likely require treatment under a bonding program. Second, and perhaps more fundamental, existing facilities where acid mine drainage treatment is occurring may cease treatment. Pennsylvania law requires operators to treat acid mine drainage permanently, and many coal operators are treating discharges as a result.[1] Because DER cannot lawfully issue a permit to an operator with violations at other sites, operators will treat existing discharges at those sites in order to obtain new permits. However, surface mining has historically been a high-turnover industry in Pennsylvania. Easily accessible high-quality coal is getting harder to find, and static coal prices have not kept pace with production costs. To the extent that operators decide to get out of the coal mining business and not obtain new permits, DER could be faced with the loss of existing acid mine drainage treatment operations. An equally significant threat stems from the actual or potential bankruptcy of existing coal operators.

In recent years, OSM has begun to give this issue serious attention, noting that "abatement and long term treatment of mine drainage continues to be the prominent environmental issue associated with mining operations" (Office of Surface Mining 1990). OSM has advised DER that acid mine drainage treatment costs must be incorporated into its alternative bonding system (letter from W. H. Tipton to Arthur A. Davis, Secretary, Pennsylvania Department of Environmental Resources, October 1991).

Several options are available to DER. These include requiring a higher up-front bond for each mining operation, requiring a higher bond at individual sites when a discharge occurs, developing a bond pool, or passing the costs along to the public through a fee or tax.

Adverse Effects of Underground Mining

More coal is produced by underground mining than by surface mining in Pennsylvania. In 1989, 40.3 million tons of coal were produced by underground mining, compared to 28.1 million tons from surface mining (Pennsylvania Coal Association 1990). Fewer surface mining operations are being permitted, and production from these operations is declining (Office of Surface Mining 1990).

Meanwhile, changes in underground mining technology—particularly the development of longwall mining—are contributing to the dominance of underground mining. Historically, coal was mined using the "room and pillar" method in which some coal was extracted, leaving rooms, and the remaining coal was left in place as pillars to provide support for mining operations. In longwall mining, by contrast, a cutting or shearing machine moves back and forth, removing large panels of coal several hundred feet wide and up to a mile or more in length. This method was not widely used until the late 1970s, after SMCRA was signed into law. However, because longwall mining requires fewer miners and because it is more efficient in removing coal (typically 70 to 80 percent, compared to 50 or 60 percent for room and pillar mining), longwall mines have become the largest and most productive of Pennsylvania's coal mines.

Underground mining has been, and continues to be, responsible for three problems—subsidence, acid mine drainage, and loss of water supplies.

Subsidence: The change from room and pillar mining to longwall mining is changing the way in which subsidence occurs. With room and pillar mining, subsidence often took years to occur, and then only in discrete areas. Indeed, DER continues to investigate subsidence complaints from mining operations that are long abandoned, and some areas have yet to subside. Longwall mining, on the other hand, results in more or less immediate subsidence over large areas. Subsidence effects include cracks or buckling in structures, ground fissures, and disruption of water supplies.

SMCRA requires mine operators to "adopt measures consistent with known technology in order to prevent subsidence causing material damage to the extent technologically and economically feasible, maximize mine stability, and maintain the value and reasonably foreseeable use of such surface lands, except in those instances where the mining technology used requires planned subsidence in a predictable and controlled manner" (sec. 516[b][l]; 30 U.S.C. 1266[b][l]). This requirement, which sounds good on first reading but which resists precise meaning under careful analysis, has been painfully difficult to implement. The hardest questions have been whether the requirement to prevent subsidence damage necessarily includes, by implication, a requirement to repair or compensate structural damage that does occur, and whether the exception for planned subsidence categorically applies to longwall mining.

OSM's original regulations in 1979 required compensation for or restoration of structures that were damaged by subsidence. Pennsylvania obtained primacy in 1982 based largely on its 1966 Bituminous Mine Subsidence and Land Conservation Act. The act prohibits mining that causes subsidence to three types of structures that were in place on April 17, 1966—public buildings and noncommercial buildings generally used by the public, dwellings used for human habitation, and cemeteries (52 P.S. secs. 1406.1–.21). DER has enforced that prohibition by requiring 50 percent of the coal under these structures to be left in place. Owners of structures built after that date can buy mine subsidence insurance from the state or buy the underlying coal from the operator. If asked, the operator is legally obligated to sell the underlying coal, with the price subject to arbitration.

The OSM rule was amended in 1983 to require protection of structures only to the extent required by state law. In 1984, a court held the amendment to be invalid because OSM had not adequately explained its departure from the prior rule. In 1985, OSM again asked Pennsylvania to require repair or compensation for material subsidence damage to any structure, including structures built before April 1966. OSM then reproposed the 1983 amendment.[2]

In 1987, OSM issued a final regulation protecting structures only to the extent required by state law and withdrew its requirement that Pennsylvania amend its program. In early 1990, a federal district court ruled that the 1987 regulation was invalid, holding that SMCRA requires repair or compensation of structures, regardless of state law limitations. OSM then issued Pennsylvania a notice of proposed preemption.

On March 22, 1991, the Court of Appeals reversed the federal district court's decision (National Wildlife Federation v. Lujan, 928 F.2d 453 [D.C. Circuit 1991]). After describing the "language, structure, and punctuation" of SMCRA as "unclear," the court held that OSM could have decided the repair or compensation question either way. OSM explained its decision not to require repair or compensation of structures by stating that the environmental interest in protection of land was different than the private contractual relationship between the landowner (and his or her predecessors) and the coal company, which could be determined under state law. Since the court found OSM's explanation adequate, it upheld the regulation.

OSM's regulation and Pennsylvania's subsidence law put all affected parties in a difficult position. Landowners and environmental interests continue to believe that longwall mining is creating significant problems. Deep mine operators, on the other hand, continue to be frustrated by the limitations on mining under pre-1966 structures. A decade of flip-flopping OSM regulation and subsequent court decisions on this question have complicated—perhaps prevented—resolution in Pennsylvania of these underlying issues.

Acid mine drainage and hydrologic balance: SMCRA also requires underground mining operators to minimize disturbances to the prevailing hydrologic balance (sec. 516[b][9]; 30 U.S.C. 1266[b][c]), but it does not require the replacement of private water supplies that are adversely affected by underground mining. The Pennsylvania program tracks the federal program by requiring underground operations to prevent subsidence damage to aquifers, perennial streams (streams that flow twelve months per year), and bodies of water that serve as significant sources for public water supplies. And it also requires underground mines to be operated to maintain the value and reasonably foreseeable uses of perennial streams as they existed before mining (compare 25 Pa. Admin. Code secs. 89.143[b] and [d] to 30 C.F.R. Sec. 817.121). But the full meaning of the hydrologic balance requirement is still being worked out.

More than 2,200 miles of Pennsylvania streams are polluted by acid mine drainage. Abandoned underground coal mines are responsible for most of these problems. Historically, outside portals were often developed at a low point in the coal reserve, so that water entering the working areas would drain to the outside. Because of the basic chemistry of coal and coal overburden, much of this water became heavily contaminated with acid, iron, manganese, and other contaminants.

DER is looking more closely at potential subsidence impacts on shallow aquifers, perennial streams, and public water supplies. DER has restricted mining underneath many perennial streams and has taken other measures to prevent surface and groundwater diminution. Problems still occur, however. In early 1991, DER was litigating a case in which longwall mining under a perennial stream severely reduced its flow. DER uses information gathered from such experiences to develop and amend its permitting program.

Because underground mines generally pump and treat water that infiltrates into them during their active phase, DER is attempting to prevent the generation of acid mine drainage at underground mines after they close. That result is commonly sought through planned layout of the underground workings. The idea is to design the workings so that they will flood upon closure, thereby limiting the amount of oxygen that can react with acid-forming materials. DER carefully reviews these plans to assure that resultant mine pools will not create artesian discharges at the land surface. Other methods include improved mine seals to prevent surface water discharges. On this issue, DER is clearly on a learning curve. Many of these methods, and the hydrologic balance protections they are intended to implement, have been used less than a decade. Although the statutory requirement to protect the hydrologic balance is vague, DER has developed policies that interpret that requirement more precisely. A significant problem in evaluating the effectiveness of new methods and policies is the life of these mines (15 to 20 years, compared with 3 to 5 years for surface mines) and their size (often several thousand acres, compared to a few hundred acres for many surface mines). Although DER will not issue underground mining permits where there will be surface discharges of acid mine drainage, including permits with gravity discharges to areas outside the mine, many mines with actual or likely discharges were already being operated when DER obtained primacy. These operators were required to post bond in an amount that would at least partially cover the cost of constructing an acid mine drainage treatment plant.

Water supply loss: SMCRA clearly and unmistakably requires the replacement of water supplies adversely affected by surface mining. On two separate occasions, the courts have ruled that SMCRA does not require replacement of private water supplies disrupted by underground mining (McElfish and Beier 1990).

The absence of explicit private water supply replacement requirements for underground mining under SMCRA or Pennsylvania law is a source of problems. DER began tracking water supply loss complaints from underground mining during the summer of 1990 and has verified a significant number of these complaints. Few things are more disruptive to a household or farm than water loss. Many underground mining companies replace damaged water supplies as a good-neighbor policy, but many do not.

A deep mine mediation project was initiated in 1986 in an attempt to create a consensus among underground coal operators, farmers, and environmental and citizen groups concerning subsidence and water resource issues. Participants were all brought together by dissatisfaction with the existing deep mining laws, although some environmental groups pulled out before the project was completed. The project's March 1990 report recommended legislative changes to provide water supply replacement for homeowners and farmers. The report also recommended requiring coal operators to restore, replace, or compensate damage to publicly owned buildings, to those used for public purposes, to residential dwellings, and to most agricultural structures. This requirement would be applicable to those structures whether they were built before or after 1966. Finally, the report recommended removal of the requirement that pre-1966 structures be protected against subsidence damage (Deep Mine Mediation Project 1990).

Legislation was introduced to effectuate the recommendations of the project. The legislature's decision will determine the environmental impacts of future underground coal mining as well as the future of Pennsylvania coal mining.

CONCLUSION

It is widely recognized that state programs provide much of the experience and thinking on which federal programs are based. It is also true that the states operate as laboratories for the effectiveness of federal programs. The substantive requirements of SMCRA were based to a large extent on Pennsylvania's program, and SMCRA in turn improved the administration and enforcement of that program. For all that, Pennsylvania's experience through the 1980s more closely resembles programmed implementation than adaptive implementation.

The adaptive implementation issues discussed in this chapter raise the old question about the appropriateness of state discretion. DER's efforts to address some of these questions, particularly the scope, meaning, and effectiveness of the hydrologic balance requirements for underground mining, clearly involve on-the-job learning over time. The decision to regulate underground mining is, among other things, a decision not to ban it. This means acceptance of some level of adverse impacts because underground mining adversely affects aquifers, whether or not structures or water supplies are damaged, and whether or not acid mine drainage or stream dewatering occurs. Pennsylvania has imposed some additional limits on underground mining; it is too soon, however, to tell whether this effort will be effective, much less how effectiveness will eventually be understood.

Apart from the generality of the statutory language, subsidence control efforts have been complicated by the technical revolution that longwall mining has brought to underground mining. Longwall mining, which gained prominence after SMCRA was adopted, significantly changes the way in which underground mining must be regulated. The requirement under Pennsylvania's 1966 subsidence act to leave 50 percent of the coal in place under certain protected structures is anachronistic in certain respects. Longwall mining also challenges the continuing economic competitiveness of surface mining. All of that weighs in favor of adaptive implementation. On the other hand, OSM's inability or unwillingness to put in place a meaningful subsidence rule has complicated resolution of this issue in Pennsylvania. The Pennsylvania legislature is likely to address this issue in the near future.

The protracted nature of the bonding issue, particularly alternative bond systems and bonds for acid mine drainage, is less explainable by the need for information or by technological developments, although more is certainly known about reclamation costs now than several years ago. The real issue seems to be concern that an adequate bonding system, if designed improperly, could impose significant and unnecessary costs on the mining industry. As a result, relatively minor adjustments to the system are made over time in the hope they will work, and their failure leads to a search for other adjustments. OSM was not active on this issue for most of the 1980s. OSM's recent interest in bond adequacy may lead to more significant changes, but it is too early to tell.

One issue, SMCRA's failure to protect water supplies from underground mining, is difficult to explain as coherent policy, particularly because water supplies are protected from the adverse effects of surface mining. In strict terms, there is nothing for Pennsylvania to implement. The debate over this issue has been made part of the overall Pennsylvania discussion about underground mining, and it may be addressed by the legislature.

It is too early to tell, then, whether adaptive implementation will eventually solve some of SMCRA's most nagging and difficult problems. But it is not too early to suggest some ways in which these problems might be addressed more quickly and effectively.

The quality of the debate over state roles is much affected by history, particularly the circumstances that led to federal regulation in the first place, and more recently by actions of the Reagan administration. State flexibility is frequently understood as inconsistent with environmental protection. Without any doubt, the cooperative federalism environmental programs developed by Congress over the past two decades, including SMCRA, have improved the effectiveness of state programs, despite (or because of) many of the coercive or nearly coercive features they contain. But an important issue, as we enter the 1990s, is how to build the capacity of states to protect the environment. A more equal mix of coercive and noncoercive features would assist that capacity building in significant ways.

OSM is required by SMCRA to assist the states in program implementation. OSM should therefore play a more active oversight role earlier on important questions such as bonding. OSM's recent guidance on the financial solvency of alternative bonding systems and its recent attention to bonding issues represent a late entrance to a play that is well under way. In addition, OSM should maintain basic program consistency over time. Although time and experience will suggest changes to any set of regulations, a zigzagging regulation is difficult to implement. As a partner in the implementation of this program, OSM gets to share credit as well as blame.

Beyond this type of basic implementation activity, Congress and OSM would do well to consider actions that would enhance the capacity of states to use discretion on behalf of environmental protection. Such actions are needed because of the increasing complexity of environmental programs in general, the increasing relative importance of state programs, the increasing sophistication, professional integrity, and environmental protection orientation of state programs, the increasing inability of Washington to dictate credible rules in large areas of environmental protection, and the continuing and growing importance of nationwide programs. A list of such actions would be lengthy, but three examples are illustrative.

First, OSM should systematically support or sponsor technical forums or conferences to enable states to share their experience with one another, particularly concerning such issues as hydrologic balance protection, where a learning curve exists. There are relatively few opportunities for state agency technical staff to talk with each other about what works and what hasn't worked, and one state's experience usually remains in that state. The proceedings from such forums or conferences should be published. To the extent that a consensus develops on individual issues, OSM might apply that consensus in program development and oversight.

Second, OSM should conduct or sponsor basic research on the environmental effects of coal mining and coal mining regulation. A balanced and full understanding of the effects of longwall mining would likely result from this research. A continuous weakness in the program is the lack of such information. Such research could greatly assist and focus program implementation.

Third, Congress and OSM need to consider ways of building state regulatory/technical assistance infrastructure, particularly with personnel. Complex environmental programs at the state level require a system that will attract and keep some of the most talented people our society produces, particularly given the importance we attach to these problems. In many states, though not to any marked extent in the Pennsylvania mining program, the most talented and ambitious people leave for the private sector after a few years. The net impact of their departure is not good for good environmental programs. As environmental programs, including SMCRA, become more complex, moreover, it is increasingly important to find new ways to hire and keep good people.

Through its first decade, SMCRA's effectiveness has been due largely to the limits it has imposed on state discretion. In the long run, however, creating conditions conducive to the environmentally responsible exercise of discretion may be at least as important to its continued effectiveness.

NOTES

1. Pennsylvania acid mine drainage law is based to a large extent on two landmark cases: Commonwealth v. Barnes and Tucker Co., 455 Pa. 393, 319 A.2d 871 (1974) (Barnes and Tucker I); Commonwealth v. Barnes and Tucker Co., 472 Pa. 115, 371 A.2d 461 (Barnes and Tucker II), appeal dismissed, 434 U.S. 807 (1977).

2. Pennsylvania successfully defended its subsidence act, in the 1986 case of Keystone Bituminous Coal Association versus DeBenedictis (480 U.S. 470 [1986]), against claims that the act violated the taking clause of the United States Constitution. The United States Supreme Court held that the act had a clear public purpose of providing for the conservation of surface land areas. In addition, it did not appear from the record that the legislation made it impossible for coal companies to profitably engage in the business, or that there had been undue interference with their investment-backed expectations.

The subsidence act frustrated the coal industry because, in its view, the act did not take into consideration the development of longwall mining. The requirement that pre-1966 structures be protected against subsidence damage is a significant inhibition against longwall mining because it is difficult to design the longwall process around blocks of coal underlying individual structures. At the same time, subsidence damage to structures following longwall mining has begun to draw a significant number of complaints.

REFERENCES

Beier, Ann E., and James M. McElfish. 1989. "The Need for Federal Oversight of Alternative Bond Systems Under SMCRA." *Policy Studies Review* 9: 132.

Berman, Paul. 1980. "Thinking About Programmed and Adaptive Implementation: Matching Strategies to Situations." In H. M. Ingram and D. E. Mann (eds.), *Why Policies Succeed or Fail*. Beverly Hills, Calif.: Sage Publications.

Costain, Ann N. 1979. "Eliminating Sex Discrimination in Education: Lobbying for Implementation of Title IX." *Policy Studies Journal* 7: 189.

Davis, Kenneth C. 1971. *Discretionary Justice*. Urbana: University of Illinois Press.

Deep Mine Mediation Project. 1990. *Consensus Proposal*.

Dernbach, John. 1986. "Pennsylvania's Implementation of the Surface Mining Control and Reclamation Act: An Assessment of How 'Cooperative Federalism' Can Make State Regulatory Programs More Effective." *University of Michigan Journal of Law Reform* 19: 903.

Hays, Samuel. 1987. *Beauty, Health and Permanence*. Cambridge: Cambridge University Press.

Kirst, Michael, and Richard Jung. 1982. "The Utility of a Longitudinal Approach in Assessing Implementation: A Thirteen-Year View of Title I, ESEA." In Walter Williams and others, *Studying Implementation: Methodological and Administrative Issues*. Chatham, N.J: Chatham Publishers, Inc.

Lindblom, Charles. 1968. *The Policy Making Process*. Englewood Cliffs, N.J.: Prentice-Hall.

Lowi, Theodore J. 1979. *The End of Liberalism*. 2d ed. New York: W. W. Norton and Co.

McElfish, James, and Ann Beier. 1990. *Environmental Regulation of Coal Mining*. Washington, D.C.: Environmental Law Institute.

Mazmanian, Daniel A., and Paul A. Sabatier. 1983. *Implementation and Public Policy*. Glenview, Ill.: Scott, Foresman and Co.

Nakamura, Robert T., and Frank Smallwood. 1980. *The Politics of Policy Implementation*. New York: St. Martin's Press.

Office of Surface Mining. 1991. *Evaluation Report: Pennsylvania Bond Program as an Alternative Bonding System*.

———. 1990. *1990 Annual Evaluation Report for the Regulatory and Abandoned Mine Land Reclamation Programs under the Surface Mining Control and Reclamation Act of 1977 in the State of Pennsylvania for the Period July 1, 1989 through June 30, 1990*.

Pennsylvania Coal Association. 1990. *Pennsylvania Coal Data 1990*.

Pennsylvania Department of Environmental Resources. 1989. *Mine Drainage Inventory and Discharge Characteristics of the Pennsylvania Bituminous Coal Region*.

Rochow, James. 1978. "The Far Side of Paradox: State Regulation of the Environmental Effects of Coal Mining." *West Virginia Law Review* 81: 559.

Udall, Morris. 1977. *Congressional Record* 123: 12,869, 12,872.

Williams, Walter. 1982. "The Study of Implementation: An Overview." In Walter Williams and others, *Studying Implementation: Methodological and Administrative Issues*. Chatham, N.J: Chatham Publishers, Inc.

10 West of the 100th Meridian: Implementation of the Surface Mining Control and Reclamation Act in the Western States

Denise A. Dragoo

INTRODUCTION

The American West has an arid climate and other unique physical conditions that exist nowhere else in the country. Thus, Congress enacted certain provisions of the Surface Mining Control and Reclamation Act (SMCRA) that apply exclusively to land west of the one hundredth meridian west longitude (the 100th meridian). These provisions include prohibiting mining in alluvial valley floors (SMCRA section 510[b][5]; 30 U.S.C. section 1260[b][5]), making exceptions to lands unsuitable prohibitions on national forest lands that lack significant forest cover (SMCRA section 522[e][2][B]; 30 U.S.C. section 1272), and establishing a separate regulatory program for special bituminous coal surface mines (SMCRA section 527; 30 U.S.C. section 1277). In addition, the vast holdings of public lands in the West required Congress to enact a federal lands program (SMCRA section 523; 30 U.S.C. section 1273) and incorporate SMCRA into federal coal leases, contracts, and permits (SMCRA section 523[b]; 30 U.S.C. section 1273).

Other than these limited provisions, Congress did not attempt to address variations in western climate, topography, and land use. Rather, it envisioned that each state program would address the "diversity in terrain, climate, biologic, chemical, and other physical conditions" (SMCRA section 101[f]; 30 U.S.C. section 1201[f]). After their programs were initially approved, many western states sought to refine their regulations to accommodate local conditions. Unfortunately, few state variations have survived program review and approval by the federal Office of Surface Mining (OSM). Federal guidance documents concerning review of state program amendments espouse the need to accommodate unique conditions. However, western states have found that OSM's review process is lengthy and demanding and that OSM prefers a uniform national standard.

This chapter will review the implementation of state programs west of the 100th meridian and the efforts of western states to accommodate their programs to national environmental standards, federal land managers, and OSM oversight. It is the thesis of this chapter that OSM has failed to allow the states adequate flexibility in administering state programs to accommodate the unique climate, biology, and terrain of the West.

HISTORY OF STATE PROGRAM IMPLEMENTATION IN THE WEST

Prior to passage of the federal Surface Mining Control and Reclamation Act in 1977, 38 states had programs to regulate coal mining and reclamation (Barry 1982). As of 1977, 10 states west of the 100th meridian regulated the surface effects of mining; these were Colorado, Kansas, Missouri, Montana, New Mexico, North Dakota, Oklahoma, Texas, Utah, and Wyoming (for a description of state programs predating SMCRA, see Hunt 1983).

Initially, western states in general favored the enactment of SMCRA. The majority of western states had laws in place prior to 1977 that required mining permits, set performance standards, and required reclamation bonds. Thus, the basic elements to regulate the coal industry were in place. Further, the coal industry in the West is typified by large, well capitalized operators, as opposed to greater numbers of small operators in the East. In contrast to the eastern coal mining industry with its longer history of mining and its numerous labor-intensive underground mines, western operations are generally more modern and larger in scale (Barry 1982). Thus, state officials and industry leaders did not expect the shift to federal standards in western states to be as difficult or costly as it was in the East (Barry 1982).

These differences between the eastern and western coal industry are well documented. For example, in 1989, 87 percent of the coal industry was employed east of the Mississippi with the majority employed in underground mines (National Coal Association 1990). By contrast, the twenty top-producing mines in 1989 were all surface operations and were located west of the Mississippi River. In states such as Wyoming and Montana, a few surface mining operations held by large companies are responsible for the majority of coal production. Wyoming alone claims four of the top five producing coal mines in the nation, which are operated by prominent oil, gas, and mining interests. The Rosebud Mine in Montana was ranked third based on 1989 production (National Coal Association 1990).

Despite the hopes of western states and industry leaders, SMCRA was not meshed into existing western reclamation programs. Rather, state reclamation programs were changed to reflect SMCRA and OSM regulatory requirements for state programs. After SMCRA was enacted in 1977, Title V performance standards were implemented in two phases. Under the first phase, states with existing reclamation laws continued to issue permits and regulate programs that incorporated interim federal standards. OSM participated with the states in joint federal/state enforcement (SMCRA sections 501, 502; 30 U.S.C. 1251, 1252). Following the interim program, states could apply to assume primary regulatory authority over state lands and cooperative authority on federal lands (SMCRA sections 503, 523; 30 U.S.C. 1253, 1273). The western states with reclamation laws all participated in the interim program and enacted enabling legislation to assume regulatory authority under the permanent program phase (SMCRA section 502[c]; 30 U.S.C. 1252; Barry 1982).

On its face, SMCRA's permanent program is administered through cooperative federalism. This term envisions a balance between federal leadership and state autonomy (Harris 1989). The principles of cooperative federalism have been broadly applied by both the Reagan and Bush administrations under Executive Order 12612, adopted October 26, 1987, and reconfirmed on February 16, 1990. Under this Executive Order, federal agencies are to accord states maximum discretion in implementing national programs administered by the states. Similarly, SMCRA reflects congressional intent that "the *primary* governmental responsibility for developing, authorizing, issuing and enforcing regulations for surface mining and reclamation operations subject to this Act *should rest with the states*" (SMCRA, section 101[f]; 30 U.S.C. sec. 1201[f]; emphasis added). Under SMCRA section 503(a), states may regulate state and private lands by enacting reclamation laws "no less stringent than" SMCRA and demonstrating that they have adequate financing and personnel to administer a program. OSM reviews state programs to be sure they are consistent with SMCRA and with federal regulations (SMCRA section 503[a]; 30 C.F.R. section 730.5 [1989]).

States that have approved programs may negotiate a cooperative agreement with OSM to regulate aspects of the federal lands program. The secretary of interior, however, retains ultimate authority to approve mine plans on federal lands and to designate federal lands as unsuitable for mining (SMCRA section 523[a]; 30 U.S.C. 1273[a]). OSM administers its own federal program in states that do not have regulatory authority and on Indian lands (SMCRA sections 504, 710; 30 U.S.C. 1254, 1300).

Unfortunately, from the western perspective, implementation of cooperative federalism under SMCRA has been less than successful (Nielson 1991). Congress provided that states should have flexibility to adopt standards to accommodate "the diversity of terrain, climate, biologic, chemical and other physical conditions" (SMCRA section 101[f]; 30 U.S.C. section 1201[f]; see generally, Barry 1982). However, rather than providing a framework for state programs, OSM's permanent regulatory program, published on March 13, 1979, was voluminous (more than five hundred pages) and detailed (44 *Federal Register* 14901; 44 *Federal Register* 15311 [1979]). These regulations were viewed as setting "uniform national rules specifying exactly how all states must proceed to comply with the standards of the Act" (U.S. Congress 1979; quoted in Barry 1982, 193). The regulations appeared to reflect the view held by OSM and environmental groups that states were not capable of regulating their local coal mining industries without a detailed blueprint.

OSM has not allowed states much latitude to adopt and enforce standards different from the uniform federal regulations. This lack of flexibility is illustrated by the so-called state window regulations that governed initial submission and approval of state programs. These regulations allowed state laws and/or regulations to vary from federal standards only if the state alternative was no less stringent than SMCRA and was necessitated by local conditions. However, OSM scrutinized state window submissions with such severity that few were approved (Barry 1982). This lack of flexibility was particularly discouraging to states that had strong reclamation laws and standards in place prior to the enactment of SMCRA.

Similarly, since implementation of their programs, western states have had difficulty in obtaining OSM approval of amendments designed to accommodate local conditions (30 C.F.R. section 732.17 [1989]). For example, Montana's and Wyoming's proposed program amendments concerning highwalls, hydrology, topsoils, and other issues unique to the West have been rejected by OSM in favor of the uniform federal standard (Montana Program Amendments, 55 *Federal Register* 19727 [May 11, 1990]; Wyoming Program Amendments, 54 *Federal Register* 52958 [December 16, 1989]).

OSM's rigidity has led to litigation in an effort to make federal regulations more responsive to local conditions in the West. For instance, in In re Permanent Surface Mining Regulation Litigation II, round II, 22 ERC 2153, 2192 (D.D.C. July 15, 1985) Judge Thomas A. Flannery remanded 30 C.F.R. section 816.46(b)(2) and section 817.46(b)(2) for OSM's failure to consider the negative impact of siltation structures in the West. In response to litigation, OSM has recently proposed a regulation to address concerns about siltation structures in the West. On November 13, 1990 (55 *Federal Register* 47430), OSM proposed a rule that would allow alternative sediment control measures for areas that receive less than 26 inches of rain annually. Unfortunately, rather than relying on the states to approve these measures, OSM is proposing to approve alternatives, based on information provided by the states. A joint committee of the National Coal Association and the American Mining Congress has requested that this proposal be withdrawn. The committee suggests that states currently have authority to approve or disapprove alternative sediment control technology and should not be required to submit evidence to OSM to support alternatives (Bureau of National Affairs 1991). In sum, OSM's lack of flexibility in oversight and state program approval has often resulted in application of uniform federal regulations. This state of affairs has been referred to by Fred Banta, former director of the Colorado state program, as "McFederalism" (Nielson 1991). This national uniformity erodes the ability of states to address the diversity of conditions unique to the West as Congress intended pursuant to SMCRA section 101(f).

CURRENT STATE/OSM RELATIONS

Structure of OSM in the West

Until October 1990, the OSM Western Technical Center in Denver, Colorado (Region V), was responsible for the oversight of permitting and inspection/enforcement matters for 14 state programs in the states of Wyoming, Utah, Louisiana, Texas, Oklahoma, New Mexico, North Dakota, Montana, Missouri, Kansas, Iowa, Colorado, Arkansas, and Alaska and for federal programs for the states of Washington and California and the Hopi, Crow, and Navajo Indian tribes. In a move to decentralize decision making, OSM established field offices in the West, located in Kansas City, Missouri, Albuquerque, New Mexico, Tulsa, Oklahoma, and Casper, Wyoming.

Despite OSM's creation of regional offices and field offices, western states have consistently viewed the agency as seeking to centralize power in Washington, D.C. Initially, OSM was organized into five regional offices throughout the United States. Region V in Denver, Colorado, was responsible for oversight of state programs in the western states and implementation of the federal program. The gravitation of decision-making power away from Region V to headquarters was evidenced during state program approval in 1979. For example, after the Montana state program was approved by Region V, OSM headquarters intervened and required the state to address more than 75 additional issues (Barry 1982).

In 1981, an effort was made to decentralize authority by creating field offices in four western states and designating the Region V office in Denver, Colorado, as the Western Technical Center. However, decisions in the regional and field offices were often overridden by headquarters. For example, the decision on Wyoming's state program amendment concerning bluffs and highwalls was pulled from Region V to Washington, D.C. (54 *Federal Register* 52958, December 26, 1989; interview, Doug Larsen, Western Interstate Energy Board [WIEB], September 20, 1990).[1] Similarly, negotiations involving a reclamation agreement and a bond for a bankrupt mine on federal lands in Utah were finalized with the state of Utah and OSM in Washington, D.C., rather than with Region V (April 13, 1990, Reclamation Agreement, Horse Canyon Mine, ACT/007/013, on file with the Utah Division of Oil, Gas and Mining, Salt Lake City, Utah).

In an effort to overcome these problems, OSM once again reorganized in October 1990. OSM has removed state oversight and inspection/enforcement from Denver, Colorado, to the field offices and renamed the Western Technical Center as the Western Support Center. The Denver, Colorado, Western Support Center now provides administrative services, technical assistance, and program support to the field offices. The Western Support Center retains the permitting jurisdiction over the federal lands program and administers the federal abandoned mine land program. The field offices, however, no longer report at the regional level; instead, they report directly to OSM in Washington, D.C., to the assistant director of field operations (interview, Al Klein, director, OSM Western Support Center, Denver, Colorado, July 2, 1991; interview, Tom Emmett, Albuquerque Field Office, OSM, July 2, 1991). This reorganization may serve to streamline the chain of command and enable the field offices to have greater control over decisions that affect the western states within their jurisdiction.

State Program Oversight

OSM exercises oversight of states by approving state programs and amendments, providing funding to state programs, and overseeing bonding, inspection, and enforcement matters. State program amendments may arise when a state seeks a change or when OSM provides notification under 30 C.F.R. part 732 that an approved state program must be amended. Notification can result from amendment of federal regulations, new cases construing regulatory requirements, or problems identified during oversight or program review (Office of Surface Mining 1989). States routinely receive notification from OSM of program deficiencies. OSM has issued more than 35 notification letters to western states (interview, John Kunz, OSM solicitor, Denver, Colorado, July 28, 1990). Most states have been able to resolve problems raised in the part 732 notifications. However, OSM did intervene for a period of time in Oklahoma to replace the state program with direct enforcement of a federal program. Oklahoma has now corrected program deficiencies and is once again enforcing its own program (Vestal 1988).

Perhaps one of the greatest concerns states have with OSM involves the difficulty of obtaining approval for program amendments to accommodate unique western conditions. A case in point concerns the struggle over OSM regulations requiring elimination of highwalls and state proposals to retain these topographic features in the West. A highwall is an unexcavated face of exposed overburden and/or coal. In the East, a highwall is viewed as a scar from past mining activities that must be removed (54 *Federal Register* 52958, December 26, 1989). However, in the West, the highwall is more consistent with the rugged landscape and may enhance raptor habitat. The state of New Mexico has amended its program to allow highwalls where they are in keeping with premining topography such as cliffs and bluffs common to the San Juan Basin (54 *Federal Register* 86464, December 31, 1989). The states of Wyoming and Montana were unsuccessful in their efforts to retain highwalls under their programs. In May 1990, OSM disapproved Montana's program regarding highwalls because the state had failed to specify the conditions that would allow their retention (55 *Federal Register* 19727, May 11, 1990). In December 1989, OSM disapproved of Wyoming's proposed retention of highwalls as a final reclamation feature to enhance nesting habitat for raptors. OSM based its disapproval on the conflict created by the Wyoming program amendment and the national standard that requires approximate original contour and highwall elimination (54 *Federal Register* 52958, December 26, 1989). Clearly, in the latter case, state efforts to accommodate regulations to local conditions appear to have been sacrificed to national uniformity.

COOPERATIVE FEDERALISM ON FEDERAL LANDS

More than 80 percent of coal reserves west of the 100th meridian are owned or controlled by the federal government (Hunt 1983). Prior to enactment of SMCRA, many western states extended their reclamation laws to federal lands. For example, Utah's Mined Land Reclamation Act, as enacted in 1975, applied "to all lands in the State of Utah lawfully subject to its police power" (Utah Code Ann. section 40–8–20 [1953 as amended]). In addition, the federal Bureau of Land Management (BLM) and the U.S. Forest Service exercised control over mining activities on lands within their jurisdiction. When SMCRA was enacted, Congress preempted state regulation of federal lands in conflict with the federal lands program (SMCRA section 523; 30 U.S.C. 1273). States with programs approved under SMCRA may negotiate cooperative agreements with OSM to regulate certain aspects of the federal lands program (30 C.F.R. part 745 [1989]). These include Colorado (30 C.F.R. section 906.30 [1989]), Montana (30 C.F.R. section 926.30 [1989]), New Mexico (30 C.F.R. section 931.30 [1989]), Utah (30 C.F.R. section 944.30 [1989]), and Wyoming (30 C.F.R. section 950.20 [1989]). While a cooperative agreement does not vest the state with complete control over mining on federal lands (NRDC v. OSM, 94 IBLA 269 [1986]), these states participate in permit review and enforcement activities, and the federal lands program must incorporate approved state program provisions that differ from the federal regulations (SMCRA section 523[a]; 30 U.S.C. 1273[a]). The secretary of interior retains authority for final approval of mine plans on federal lands (30 C.F.R. 740.4[a] [1989]; In re Permanent Surface Mining Regulation Litigation, Surface Mining II, round I, 21 ERC 1193, 1195 [D.D.C. 1984]) and for designation of federal lands as unsuitable for mining (30 C.F.R. 740.4[c][1989]).

Federal lands within western states that have not negotiated a state-federal cooperative agreement are regulated by OSM. These include Alaska (30 C.F.R. section 902 [1989]), Arkansas (30 C.F.R. section 904 [1989]), Kansas (30 C.F.R. section 916 [1989]), and Oklahoma (30 C.F.R. section 936 [1989]). In addition, federal programs are in effect in California (30 CFR section 905.700 [1989]), Idaho (30 C.F.R. section 912.700 [1989]), and Washington (30 C.F.R. section 947.700 [1989]). OSM currently regulates Title V of SMCRA on Indian lands pursuant to the federal program, although tribes may enter into agreements with OSM to participate in the abandoned mine lands program. P.L. 100–71 amends SMCRA to authorize the Crow, Hopi, and Navajo tribes to obtain approval of abandoned mine land programs without approval of regulatory programs (30 C.F.R. part 756 [1989]). OSM also retains jurisdiction over split estate lands located off reservation but owned by an Indian Tribe (Pittsburgh and Midway Coal Mining Co. v. OSM, 115 IBLA 148 [1990]).

Under the federal lands program, OSM coordinates decisions with federal land managers. Pursuant to the Federal Land Policy and Management Act of 1976 and the Mineral Leasing Act, as amended, the Bureau of Land Management has authority to lease coal reserves on public lands (amended by the Federal Coal Leasing Amendments Act of 1976, P.L. 94–377, 90 Stat. 1082).[2] Federal coal leases and exploration permits contain stipulations that require compliance with SMCRA (section 523[b]; 30 U.S.C. section 1273[b]). Federal coal leases on national forest lands are subject to additional environmental assessment and land use stipulations imposed by the Forest Service (Federal Coal Leasing Amendments Act of 1976; P.L. 94–377, 90 Stat. 1082; 30 U.S.C. section 201[a][3][A][iii]). Federal lands are reviewed for their suitability for coal leasing under section 522 of SMCRA.

Before OSM can issue a mining permit, it must obtain approval from the federal land manager (30 C.F.R. section 740.5[3][4][1989]). Bureau of Land Management or the Forest Service review and approve mining and operation portions of proposed mine plans on public lands and forest lands, respectively. Bureau of Land Management's requirements for a resource recovery and protection plan are incorporated into the mine permit application (43 C.F.R. parts 3480–3487 [1989]; 30 C.F.R. section 740.5[d][4][1989]). Once permits are issued, coal mining activities are regulated by the state and OSM pursuant to cooperative agreements, with input from federal land management agencies. (For a discussion of the duplication in state/federal permit review initially undertaken on federal lands, see Barry 1982.) In states with cooperative agreements, OSM is cobeneficiary on reclamation bonds for mining activities on federal lands (30 C.F.R. section 740.15[b][1989]).

PROHIBITION OF MINING

Lands west of the 100th meridian are subject to unique unsuitability standards and alluvial valley floor restrictions that may result in the prohibition of coal mining activities. Before they issue a coal mining permit, state and federal regulatory authorities must determine that lands are not unsuitable for mining and are not located within an alluvial valley floor. A determination concerning alluvial valley floors is unique to lands located west of the 100th Meridian. In addition, federal lands are subject to intensive unsuitability review by the secretary of interior and federal land management agencies. Because a high percentage of federal lands are located west of the 100th meridian, the federal lands review program has a significant impact on land use in western states.

Alluvial Valley Floors

The operator has the burden under section 510 of SMCRA to establish that the application to mine is in compliance with all requirements of the state or federal program (SMCRA section 510[a]; 30 U.S.C. 1260[a]). West of the 100th meridian, operators must also show that the mine will not (1) interrupt, discontinue, or preclude farming on alluvial valley floors (AVFs), or (2) materially damage AVF water supplies (SMCRA section 510[b][5]; 30 U.S.C. 1260[b][5]). Similarly, the operator must meet environmental performance standards to preserve the essential hydrologic functions of AVFs in arid and semiarid areas (SMCRA section 515[b][10][F]; 30 U.S.C. 1265[b][10][F]). In enacting these provisions, Congress sought to preserve AVFs as "productive lands that form the backbone of the agricultural and cattle ranching economy" of the West (U.S. Congress 1977). Undeveloped range lands and small areas that have negligible impact on farming are excluded from the AVF mining ban, as are certain operations that produce coal in commercial quantities or mines permitted by states before 1977 (SMCRA section 510[b][5]; 30 U.S.C. 1260[b]5]; 30 C.F.R. section 822.12[b][1989]).

Before a permit is issued, the regulatory authority must determine whether an AVF exists within or adjacent to the permit boundaries. Field studies must be conducted to determine the presence and extent of an AVF and the hydrologic regimes that support it (30 C.F.R. section 785.19 [1989]). If an AVF is found, the regulatory authority must further determine the impact of the operation on the AVF and its water supply (SMCRA section 510[b]; 30 U.S.C. 1260[b]). A permit can be granted if the regulatory authority finds that the operation (1) would not interrupt, discontinue, or preclude AVF farming; (2) would not materially damage AVF water supply systems; and (3) would meet AVF environmental performance standards (30 C.F.R. section 785.19[e][1][1989]). If a permit is approved in an area affecting an AVF, it must meet specific design criteria, performance standards, and monitoring and reclamation requirements (30 C.F.R. section 785.19[1989]; 30 C.F.R. part 822 [1989]).

AVF requirements were adopted by Congress to address arid conditions west of the 100th meridian; however, states have had little flexibility in administering these requirements. Initially, OSM adopted detailed specifications on the information required in a permit application concerning AVF's and their water supplies. In 1983, the secretary of interior withdrew these detailed regulations to allow the regulatory authority discretion to make AVF determinations based upon diverse site-specific conditions (48 *Federal Register* 29814 [1983]). The National Wildlife Federation (NWF) challenged withdrawal of the detailed AVF regulations, and the trial court ruled in NWF's favor, remanding the secretary of interior's regulations (In re Permanent Surface Mining Regulation Litigation II, round II, 21 E.R.C. 1724, 1740 [D.D.C. 1984] [PSMRLI, round II]; aff'd NWF v. Interior Dept., 27 E.R.C. 1153, 1185 [D.C. Cir. 1988]).

OSM has once again promulgated detailed AVF regulations (30 C.F.R. section 785.19; 48 *Federal Register* 29820, June 28, 1983, as amended; 54 *Federal Register* 9735, March 7, 1989). Although these regulations are touted as guidelines, states have argued that they are essentially uniform national standards that can be applied with little or no flexibility. For instance, OSM regulations apply the AVF performance standard concerning hydrologic protection to all AVFs, whether or not they are significant to farming. These regulations were upheld in NWF versus Interior Department, 27 ERC 1153, 1200 (D.C. Cir. 1988). However the secretary of interior has recently defined farming to exclude ranching, livestock, and grazing activities (54 *Federal Register* 9724 [March 7, 1989]). This regulation was upheld in NWF versus Lujan, 1990 (U.S. Dist. LEXIS 11682, September 5, 1990), and it provides western operators some relief from the AVF mining ban.[3]

Designation of Lands Unsuitable

Lands may be designated as unsuitable for mining pursuant to section 522 of SMCRA under three separate processes (30 U.S.C. section 1272). Congress has determined that certain lands are unsuitable per se for mining under SMCRA section 522(e). In addition, Congress has set certain unsuitability criteria and has required the review of federal lands under this criteria pursuant to section 522(b). Finally, the regulatory authority must respond to a citizen petition to designate lands unsuitable under SMCRA section 522(c) (see generally, Dragoo and VanBuskirk 1982.)

Due to the predominance of federal lands west of the 100th meridian, the federal lands review process has had a significant impact on western states. SMCRA section 522(a)(5) integrates unsuitability criteria into the land use decisions of BLM, the Forest Service, and state and local governments. Federal land managers apply unsuitability criteria in comprehensive land use plans and environmental assessments on coal lease applications (43 C.F.R. section 3461.3–1[b] [1988]). For instance, in the state of Utah, more than 50 percent of all coal produced is extracted from underground mines on Forest Service lands. Unsuitability criteria have been applied in land use plans for the Manti-LaSal National Forest, the

Fish Lake National Forest, and the Uinta-Southwestern Utah Coal Region round II, Final EIS (Bureau of Land Management 1983).

The Bureau of Land Management has adopted some twenty unsuitability criteria under 43 C.F.R. section 3461.5, and these criteria are even more detailed than OSM unsuitability regulations. Although underground mining activities are generally allowed on National Forest lands west of the 100th meridian under section 522(e), proposed changes in unsuitability criteria may dramatically reduce mining. Under a proposed regulation withdrawn by OSM in June, 1989, subsidence caused from underground mining would be severely limited or disallowed in areas defined by SMCRA section 522(e), including national forests. Because this regulation would have significant impact, the state of Utah and the Utah Mining Association requested its withdrawal (Governor Norman H. Bangerter, letter to OSM, April 19, 1989; Dragoo 1990).[4]

The designation of federal lands as unsuitable is further complicated by land ownership patterns in western states. Since the early 1800s, when a western territory was admitted to statehood, Congress granted certain lands for support of public schools and other institutions. For instance, the enabling acts of Utah, New Mexico, and Arizona grant four sections within each township (Utah Enabling Act of July 16, 1894, ch. 138, section 6, 28 stat. 107, 109 [1894]; Act of June 20, 1910, ch. 310, sections 6, 24, 36 stat. 557, 561–572–573 [1910]). In Utah, these lands are generally located in sections 2, 16, 32, and 36 of each township. State lands scattered throughout the public domain are strongly influenced by federal land use policy. This land use pattern compelled Utah to file an action challenging the secretary of interior's designation of federal lands unsuitable in the Alton coal fields of southern Utah. The state challenged the designation as a "taking" of state coal reserves that were surrounded by blocks of federal lands determined unsuitable for mining because of their proximity to Bryce Canyon National Park. This action was dismissed by Utah Federal District Judge David K. Winder (Utah Int'l., Inc. v. Watt, 553 F. Supp. 872 [D. Utah 1982]). Utah has, however, proposed federal legislation to consolidate its state lands into large blocks to avoid this type of land use conflict (see Project Bold 1982; Matheson 1988).

ABANDONED MINE LANDS PROGRAM

The Abandoned Mine Lands Reclamation (AML) program is set forth at Title IV of SMCRA to provide for the reclamation of mined areas inadequately reclaimed prior to the enactment of SMCRA (SMCRA section 102[h], section 401 et seq.; 30 U.S.C. section 1202[h], 1231, et seq.). To fund reclamation projects, under 402 SMCRA, a fee is assessed of 35 cents per ton for coal produced by surface mining operations and 15 cents per ton for coal produced by underground mining (30 U.S.C. section 1232). Reclamation fees are collected by state and federal regulatory authorities and deposited in the Abandoned Mine Reclamation Fund (SMCRA section 401; 30 U.S.C. section 1231). States with approved programs may receive funds for reclamation projects approved by the secretary of interior (SMCRA section 405[c]; 30 U.S.C. section 1235[c]). A state may receive up to 50 percent of funds raised within the state. Public Law 100–71, enacted July 11, 1987, amends Title IV of SMCRA to enable the Crow, Hopi, and Navajo tribes to obtain approval of abandoned mine land programs without having approved regulatory programs (30 C.F.R. part 756 [1989]). After OSM has approved a reclamation program for a state or an Indian tribe, the state or tribe is eligible to receive reclamation grants from the Abandoned Mine Land Reclamation Fund (30 C.F.R. part 886; 30 C.F.R. part 756 [1989]).

The Western Support Center in Denver, Colorado, currently administers OSM's abandoned mine land program in the west. All western states with approved regulatory programs also have approved AML programs. The Crow, Hopi, and Navajo tribes have also qualified for AML funding.

The funding of noncoal reclamation projects is a major concern for western states and Indian tribes. Throughout the long history of coal mining in the East, numerous coal mines were abandoned before SMCRA was enacted. Coal mining in the West is characterized by more modern operations, and fewer coal mines have been abandoned in the West than in the East. Many western states have completed or nearly completed their coal reclamation projects. Montana and Wyoming have certified complete coal mine reclamation in their states. Colorado, Utah, and New Mexico are nearing completion. These western states have reclaimed more than two thousand coal mines to date and are now seeking to broaden their use of funds for noncoal projects, including abandoned mineral mines.

Currently, mineral mines and mines other than coal mines are being reclaimed and sealed under SMCRA section 409 (30 U.S.C. section 1249). This provision allows the governor of any state or the chairman of any tribe to request funding from the Abandoned Mines Land Fund. Grants may be obtained to fill voids, seal abandoned tunnels, shafts, and entryways, and reclaim the surface impacts of underground or surface mines. To date, approximately four thousand noncoal mines in the states of Colorado, Utah, Montana, New Mexico, and Wyoming have been reclaimed under SMCRA section 409. However, this is less than one-tenth of the remaining abandoned noncoal mining operations that pose hazards in these states (interview, Mary Ann Wright, president, Association of Abandoned Mine Land Programs, Salt Lake City, Utah, September 27, 1990).

The Abandoned Mine Land program, slated to expire on August 3, 1992, was reauthorized by Congress on October 26, 1990, as part of the 1991 Budget Reconciliation Act. The inequities arising from the use of abandoned mine fees assessed in the West to reclaim coal mining operations in the East were considered by proposed AML reauthorization bills (letter from Congressman Nick J. Rahall, II, chairman, Subcommittee on Mining and Natural Resources to Doug Larsen, Western Interstate Energy Board, November 16, 1989). In 1989, H.R. 2095 was proposed to extend the AML program, and it introduced a new program to reclaim noncoal mines, funded by a fee on mineral materials disposed of pursuant to section 3 of the act of July 31, 1947 (30 U.S.C. section 603). The noncoal AML program was rejected by Congress, and the AML coal program was extended for only three years, until September 30, 1995 (Title IV, section 402[b]). Although the noncoal AML program was not enacted, certain provisions were made for funding impacts related to mineral mines. Until all coal mines have been addressed, AML funds can be expended only on priority one (i.e., emergency) noncoal sites. Once states or tribes have certified that their coal programs are completed, they may fund activities or public facilities related to the coal or minerals industry (Title IV, sections 409[c] and 411[c]). In addition, the provisions of sections 407 and 408 authorizing entry of state AML personnel onto abandoned mine lands and application of liens to recoup reclamation costs have been extended to mineral operations (Title IV, sections 411[f] and [g]).

CONCLUSION

In sum, SMCRA has been implemented in the western states in a manner that favors national uniformity over unique local conditions and western state programs. Although OSM has reorganized its regional structure to streamline chain of command from western field offices to headquarters, this restructuring has done little to enhance state autonomy over programs. OSM has heavily scrutinized state window provisions and program amendments, leaving the states with little discretion to address unique local conditions. While this tight regulatory rein is not unique to western states, it does impact the West in unique ways. The detailed federal regulatory program is based largely on conditions encountered in eastern mining. Therefore, the arid and semiarid conditions and unique topographic features of the West, such as bluffs and cliffs, are not adequately addressed by either state or federal programs. The predominance of federal land ownership in the West increases the influence of federal regulators on leasing, permitting, and land-use decisions on public lands and adjacent state lands. This result is contrary to the intent of Congress to address the "diversity of terrain, climate, biologic, clinical, and other physical conditions" through state programs. The unique conditions of the West can be accommodated under SMCRA only when federal regulators are willing to give western states greater latitude to address local conditions in their state programs.

NOTES

1. The WIEB Reclamation Committee represents the states of Montana, Wyoming, North Dakota, Colorado, New Mexico, and Utah (54 *Federal Register* 52958, December 26, 1989). This issue was politicized by the involvement of Wyoming's Congressional delegation.

2. As an aside, it should be noted that the federal coal leasing program, although implemented almost entirely on western lands, is not free from East/West politics. West Virginia Representative Rahall recently introduced H.R. 693, the Federal Coal Equity Act of 1991, requiring BLM to consider the impact of a western coal lease on the production of coal east of the 100th meridian.

3. The AVF mining ban raises taking concerns under the 5th and 14th amendments of the U.S. Constitution. This issue was recognized most recently in Whitney Benefits, Inc. versus U.S. No. 90–5058, 926 F.2d 1169 (Fed. Cir.), *cert. denied*, no. 91–195 (Nov. 18, 1991), *aff'g* 18 Cl. Ct. 394 (1989).

4. The "state law limitation" on the operator's obligation to correct subsidence damage under 30 C.F.R. section 817.121(c)(2) (1989) was recently upheld in National Wildlife Federation versus Lujan, 928 F.2d 453 (D.C. Cir. 1991), March 22, 1991.

REFERENCES

Barry, Hamlet, III. 1982. "The Surface Mining Control and Reclamation Act of 1977 and the Office of Surface Mining: Moving Targets or Immovable Objects?" *Rocky Mountain Mineral Law Institute* 27A: 169–337.

Bureau of Land Management. 1983. Uinta-Southwestern Utah Coal Region, Round II, Formal Environmental Impact Statement.

Bureau of National Affairs. 1991 (March 3). *Environmental Reporter*, p. 1991.

Dragoo, Denise A. 1990. "Subsidence Controls and Section 522 of the Surface Mining Control and Reclamation Act." University of Utah, *Journal of Energy, Natural Resources and Environmental Law* 2(1).

Dragoo, Denise A., and Ronald VanBuskirk. 1982. "Designation of Coal Lands as 'Unsuitable' for Surface Coal Mining Operations." *Rocky Mountain Mineral Law Institute* 27A: 1–24.

Harris, R. 1989. "Federal-State Relations in the Implementation of Surface Mining Policy." *Policy Studies Review* 9: 1.

Hunt, Dean K. 1983. *Coal Law and Regulation* 2:40.01.

Matheson, Scott M. 1988. "Improving Public Land Management Through Land Exchange: Opportunities and Pitfalls of the Utah Experience" *Rocky Mountain Mineral Law Institute* 33: 4.

National Coal Association. 1990. *Facts About Coal 1990* 16: 21.

Nielson, Dianne R., Director, Utah Division of Oil, Gas and Mining. 1991. "Forging a Three State/Federal Relationship," American Mining Congress Convention 1991, Coal Policy Session, October 1.

Office of Surface Mining. 1989. "Directive Concerning Processing of Proposed State Regulatory Programs, Amendments and Part 732 Notifications." (Available from OSM, Western Support Center, Denver, Colorado.)

———. 1989. Oversight Reports, Fiscal Year 1989 Annual Evaluation Reports. (Available from OSM, Western Support Center, Denver, Colorado.)

Project Bold, Alternatives for Utah Land Consolidation and Exchange. 1982. Utah Department of Natural Resources.

Public Land Law Review Commission. 1970. *One Third of the Nation's Land.*

U.S. Congress. 1977. House Report No. 218, 95th Cong., 1st sess. 116, Reprinted in 1977 *U.S. Code Cong. & Admin. News* 593, 649.

———. 1979. Senate Committee on Energy and Natural Resources *Report on Surface Mining Amendment of 1979* S. Rep. No. 96—271, 96th Cong., 1st sess. 13.

Vestal, Thomas M. 1988. "First Decade of the Implementation of the Surface Mining Control and Reclamation Act in Oklahoma." *Tulsa Law Journal* 23: 593.

11 State Performance under the Surface Mining Control and Reclamation Act

Michael J. Scicchitano, David Hedge, and Patricia Metz

THE GROWTH OF PARTIAL PREEMPTION

During the 1960s and 1970s, Congress enacted a number of regulatory programs, such as the Clean Air Act and the Surface Mining Control and Reclamation Act (Advisory Commission on Intergovernmental Relations 1984) and committed billions of dollars to improve environmental quality (Lilley and Miller 1977).

Many of these regulatory laws adopted the partial-preemption approach (Advisory Commission on Intergovernmental Relations 1984), an approach that attempts to balance the demands for national involvement to improve the environment with the traditional state role. The federal government establishes national regulatory standards; then, if a state closely follows these standards, the federal government grants it the right to implement the regulatory program.

The framers of partial-preemption programs seem to have adopted the hierarchical approach as the appropriate means to implement intergovernmental regulatory programs. The degree to which the federal government accepts and follows this hierarchical view of federalism is not constant and may vary over time. In regard to the surface mining program, for example, initial federal actions were "hierarchical in nature," but the federal commitment to a hierarchical style waned considerably in the early 1980s, during the first years of the Reagan administration (Menzel 1981, 214).

It is important to note that state efforts to control implementation of regulatory programs occurred before Reagan administration efforts to relax federal controls. During the Carter administration, for example, federal officials who enforced the surface mining program responded to pressures emanating from the states (Hedge and Jallow 1990). Interstate differences in federal enforcement were related to both congressional and industry influences. States with more powerful coal industries tended to experience lower rates of federal enforcement.

It appears that bottom-up influences prevailed during the early 1980s, following the Reagan administration's efforts to deregulate and devolve greater regulatory authority to the states. Great diversity occurred in how states implemented federal standards (Thompson and Scicchitano 1985; Scicchitano, Hedge, and Metz 1989), and serious shortcomings affected the quality of state enforcement during those years (U.S. General Accounting Office 1986). Some of these deficiencies reflected the unwillingness of states to carry out federal programs. But problems in state enforcement also resulted from a lack of federal oversight and state capacity.

For many, the lackluster performance of the states simply verified the fear that devolution would produce de-facto state regulation. Moreover, others feared that once authority was delegated to the states, federal officials would be unable to retrieve that authority (Eads and Fix 1984). This pattern of failed oversight and state neglect improved by the mid-1980s. As a result of congressional and interest-group pressures, Reagan officials appeared to adopt a more balanced approach to regulation.

This chapter provides a brief overview of how federal authorities attempted to direct state regulatory behavior and then looks closely at how state officials implemented federal standards under the Surface Mining Control and Reclamation Act of 1977 (SMCRA). In particular, it traces state enforcement over time to determine the extent to which state regulatory enforcement is both diverse and responsive to federal controls.

Federal-State Roles in SMCRA Enforcement

The Surface Mining Control and Reclamation Act of 1977 was passed to correct one of the country's most glaring environmental problems, improper surface mining. The intent of Congress was, primarily, to implement a national program to improve and protect the environment by reclaiming already spoiled land and preventing improper surface-mining practices.

Under SMCRA, the Office of Surface Mining (OSM) was created to develop surface-mining regulations. Following enactment of SMCRA, OSM began to implement the state programs. States were required to submit proposals for their own regulatory programs by January 1979. Those proposals had to meet "the standards of the act and the federal permanent program" (Shover et al. 1986, 38). Once a state program was approved, OSM's role shifted to oversight—monitoring state programs to insure that they remained faithful to the federal guidelines. OSM retained the ultimate sanction of revoking primacy when states failed to adequately implement any aspect of SMCRA.

State inspectors must perform several key tasks to enforce the SMCRA properly. The first of these tasks is to inspect mines. Every year state inspectors are required to conduct four complete inspections of each mine. Inspections are complex activities during which state enforcement officials must check paper files that relate to the mine permit as well as to water monitoring and blasting. In addition, inspectors must check the mine site for conditions that relate to a range of performance measures, such as revegetation, topsoil handling, backfilling, and even the mine access road.

If inspectors detect a violation of some standard, such as failure to properly revegetate the land, then they must issue a Notice of Violation (NOV), giving the operator 90 days to develop and implement a plan to correct the violation. If the violation is not corrected within that time, the inspectors must issue a Cessation Order (CO) to stop mining until the violation is corrected. Inspectors must reinspect the mine within three days of the scheduled abatement date. Finally, operators are fined for these infractions. Penalty fees for both the NOVs and the COs allow inspectors some discretion in setting the amount. Operators, of course, have the right to appeal enforcement actions. States have some flexibility in establishing penalty amounts, but the amounts states set must be at least as large as those required by the federal program.

It is crucial for federal officials to oversee the enforcement actions of the states. Federal oversight is designed to insure that state regulatory programs do not overlook mining infractions. Lax state enforcement could mean, simply, that national program goals would not be achieved. OSM itself also conducts inspections of mine sites to insure that state enforcement agents cite all improper mining practices. If OSM inspectors detect an infraction, they can compel a state to issue a citation or provide reasons why it should not be issued.

While the federal government does have formidable tools available to direct the behavior of the states, the commitment of federal officials to use these tools is less certain (Mazmanian and Sabatier 1983; Bullock and Lamb 1984). Over time, the commitment of federal officials to insuring that the national goals of the surface mining law may change. The Carter administration, for example, had a strong commitment to insuring that the states achieved national surface mining regulations. It vigorously enforced the surface mining law in the states.

In the 1980s, however, the Reagan administration took a substantially different approach to SMCRA implementation; it wanted to reduce the federal role and increase the state role in intergovernmental affairs (Nathan and Doolittle 1984). It attempted to limit federal oversight and increase state discretion; it proposed changes in regulations that would enhance state flexibility in designing programs (Menzel 1983).

Variations in State Implementation of the SMCRA

Even when federal officials actively sought a more uniform attainment of national regulatory goals, it is unlikely that OSM could totally eliminate state discretion. A traditional argument used by opponents of devolution is that greater discretion would result in competition among states that would drive the level of regulation down (Rowland and Marz 1982; Vestal 1988). Research has demonstrated that states will not automatically reduce enforcement of partial-preemption programs as their discretion increases; some states will increase the level of enforcement while others will decrease it (Thompson and Scicchitano 1985). Other research found considerable variation in replacing cuts made by the Reagan administration in federal grants-in-aid programs (Nathan and Doolittle 1987). Under the Reagan administration, therefore, we could expect substantial differences in the level of surface mining enforcement among the states as they responded to various forces for or against regulation.

A number of political or organizational factors exist that explain why some states might implement the surface mining law more vigorously than do others (Hedge, Scicchitano, and Metz 1988). One important factor relates to differences among the states in the nature of the reclamation task (Mazmanian and Sabatier 1983). It is, simply, more difficult to reclaim surface-mined land in some states than in others. The more difficult the reclamation task, the less likely that miners will follow and states will energetically enforce surface mining regulations. The technical difficulties in mining are directly related to the three surface mining regions: the East, the Midwest, and the West (Harris 1985).

The eastern region includes the states of Alabama, Kentucky, Maryland, Pennsylvania, Tennessee, Virginia, and West Virginia. The topographical pattern of this region differs substantially from those in the other two regions. In the East, most surface mining takes place on steep, sloping terrain that presents immense problems when operators attempt to return the land to its approximate original contour. Also, the seams of coal are thinner and the mines are smaller in this region than in the other two. It was largely because of the eastern miners, "with their considerable talent for ecological and economic recklessness," that the SMCRA was passed (Rosenbaum 1987, 173). The most stringent regulatory provisions of the SMCRA were directed toward the Appalachian region (Rosenbaum 1987). When the act is being stringently enforced, we would expect a higher level of enforcement in the East than in the other regions.

The midwestern region includes Arkansas, Illinois, Indiana, Iowa, Kansas, Missouri, Ohio, and Oklahoma where surface mining is much less difficult. The mines in this region are generally larger and have thicker seams and less overburden. The western region has even thicker seams of coal and larger mines. Western land is relatively flat range, and the topography is not a problem for mining (Harris 1985, 69).

A second factor related to the ability of state efforts to implement the SMCRA is mine size. Smaller mines, typically found in the eastern mining region, are often family run, have limited resources, and do not have the best record of complying with the SMCRA (Harris 1985). Many of the larger corporate-owned mines in the Midwest and West typically employ specialists to insure that the provisions of the surface mining law are followed. It is, clearly, more difficult for operators of smaller mines, particularly those in the East with their history of improper reclamation practices, to change their behavior to comply with the SMCRA (Mazmanian and Sabatier 1983, 24). Thus, we expect states in the eastern region to be least likely to comply with the SMCRA, both on the basis of the technical difficulty of mining and the behavioral change required.

Implementation and Surface Mining Enforcement

A previous analysis of the state response to devolution examined how, during the 1980s, the top-down influence of the federal sovereigns, as well as the bottom-up influences within the various mining regions, affected state implementation of the SMCRA (Scicchitano, Hedge, and Metz 1989). The eastern region, for example, was most affected by the efforts of the Reagan administration to decentralize enforcement behavior. In the early 1980s, when the states received primacy and assumed responsibility from the federal government for implementation of the SMCRA, substantial differences appeared in the enforcement behavior of the West/Midwest and the East. After they received primacy, states in the West and Midwest increased enforcement for all three measures. States in the eastern region did increase enforcement activity for NOVs, but at a substantially lower level than those in the other two regions. Most importantly, the eastern states actually reduced enforcement of cessation orders, which are probably the most important of the enforcement measures.

Table 11.1
State Enforcement Activity in 1989 (By Region)

Region	Inspections	Notices of Violation	Citations
East	4.56	.51	.07
Midwest	4.04	.65	.08
West	2.79	.50	.01

Source: 1989 State SMCRA Annual Evaluation and OSMRE Annual Reports.

All data adjusted for the number of inspectable units.

By mid-decade, the NOV and CO enforcement actions of the East remained lower than those of the West and Midwest. The level of eastern NOVs was about half that in the other two regions (Scicchitano, Hedge, and Metz 1989). It seems clear, therefore, that when pressures from the federal government eased, the eastern states were not as willing to implement the SMCRA as were the western and midwestern states.

By the mid-1980s federal officials increasingly emphasized that OSM should promote more vigorous implementation of the SMCRA among the states. Most of these efforts to improve enforcement seemed to be directed toward the East. Levels of enforcement for inspections, NOVs, and COs increased more rapidly in the East than in the Midwest or West. In 1989 we can see the impact of federal efforts to increase state enforcement. The mean regional levels of enforcement for state inspections, NOVs and COs are presented in Table 11.1. Because the levels of enforcement would be affected by the amount of mining activity, the figures in Table 11.1 have been adjusted for a per-inspectable-unit (mining and other coal-related facilities) basis.

The levels of state enforcement of the SMCRA reached some level of equilibrium by 1989. The states appear to have come to some understanding of appropriate levels of enforcing the surface mining law. There were no longer great disparities in the levels of SMCRA enforcement among the mining regions. The similarity is striking between the eastern and midwestern regions where there are only slight differences in enforcement activities. Figures for the western region are somewhat lower, especially the number of inspections. From the perspective of the basic enforcement measures, it appears that the renewed commitment of federal officials influenced the states and promoted greater levels of enforcement.

THE QUALITY AND EFFECTIVENESS OF STATE ENFORCEMENT

The enforcement measures focused on thus far—inspections, NOVs, and COs—provide only a limited view of SMCRA implementation. To better understand the dynamic relationship of federal top-down and state bottom-up influences, we need to move beyond these numbers and examine the quality of enforcement—the degree to which all existing infractions are detected and cited. We will also determine how effective the SMCRA is likely to be in encouraging coal operators to improve mining practices. To assess the quality and potential effectiveness of state enforcement, we will use data from the annual state evaluation and OSM reports.

Penalties and the Effectiveness of State Enforcement

It is difficult to measure the potential effectiveness of state enforcement activities in promoting environmentally sound surface mining practices. But among the best indicators of potential effectiveness is the amount of penalties states assess and then collect from mine operators. Penalties are assessed when inspectors identify an improper mining practice and issue a NOV. State programs set minimum penalties and inspectors have some discretion in specifying the amount of a penalty. In 1986, for example, the Pennsylvania program had a minimum penalty of $750 per day for a violation that continued beyond the abatement period, and a fine of $2,000 per acre for removing topsoil or overburden from an unpermitted area. The larger the amount of penalty, however, the greater the financial harm to the miner and the greater incentive to improve mining practices.

As can easily be imagined, mine operators are not often eager to pay these penalties. The actual amount collected, as an indicator of the hardship imposed on mine operators, may be a better measure of the willingness of the states to improve mining conditions than the amount assessed. In this analysis, therefore, we examine both the amount of penalty assessments and collections. The ultimate responsibility for collecting outstanding assessments rests with the state programs and state legal systems. OSM has attempted to assist the states, most notably by developing a computer system for tracking mine operators with outstanding fines.

An examination of penalty actions from the middle 1980s to the later 1980s will help us assess the changing commitment of federal sovereigns to improving state implementation of SMCRA. We present a summary of penalty information by region. Because penalties are primarily generated

Table 11.2
State Assessment and Collection of SMCRA Penalties (By Mining Region)

Region	Assessments 1986	Assessments 1988	Collections 1986	Collections 1988
East	$981.89	$212.30	$96.60	$118.08
Midwest	$783.74	$299.40	$699.94	$141.81
West	$426.37	$236.97	$539.98	$245.75

Source: State SMCRA Annual Evaluation and OSMRE Annual Reports.

All data adjusted for the number of NOVs issued.

by NOVs, all data in this analysis are standardized by the number of NOVs. Data on penalty amounts were obtained from the annual state reports on SMCRA enforcement activities.

Data in Table 11.2 show a substantial difference between the eastern and midwestern/western regions in penalty activity for 1986. The East had both the highest level of penalty assessments and the lowest level of collections of any region. While the East was actively issuing large assessments, it was ineffective in collecting those fines. Mine operators likely would not be motivated to improve their mining operations if the fines they were assessed were not collected. In contrast, the West and Midwest both had substantial levels of penalty assessments and equally substantial, if somewhat smaller, levels of collections.

By 1988, the nature of penalty activity had changed. In all regions, the level of assessments declined from 1986 to 1988, with the largest decline in the East, where assessments fell by more than $700 per NOV. While collection of penalties increased in the East, it declined in the other regions. Thus, there was less difference between the eastern region and the other regions in the amounts of both penalty assessments and collections. Mine operators who received assessments would be more certain that states would collect those fines. It is not clear, however, from Table 11.2 why penalty amounts declined in all regions. The data cannot be explained by a decline in the number of inspectable units from 1986 to 1988 (only about 4 percent) nor by a reduction in the number of complete inspections (which remained about the same during this period). It may simply be that miners were learning how to comply with the provisions of the surface mining law. Further inquiry is needed, but possibly by the end of the 1980s mining practices had improved, and that improvement is reflected in the smaller levels of assessments.

Table 11.3
Citizen Complaints of Mining Activity (By Mining Region)

Region	Complaints 1986	Complaints 1989
East	.23	.17
Midwest	.13	.09
West	.11	.08

Source: State Annual SMCRA Evaluation and OSMRE Annual Reports.

All data adjusted for the number of inspectable units.

The Quality of State Enforcement

The quality of state enforcement relates to the degree to which state enforcement agents cite all existing SMCRA infractions. Again, while levels of enforcement among regions were similar by 1989, because of the difficulties of mining in the East levels of eastern enforcement should be higher. Clearly, it is difficult to measure the number of SMCRA infractions that state inspectors might overlook. In this section we use two measures to assess the quality of state enforcement: the number of citizen complaints about mining practices and the number of improper mining practices identified by federal inspectors who oversee state enforcement activities.

Under the SMCRA, citizens may file formal complaints with the state enforcement agency regarding mining practices they believe are improper. State enforcement agencies are expected to investigate the situation and, if necessary, take enforcement action. It is reasonable that the number of citizen complaints would be higher in those states that do the least effective job of enforcement. In this analysis, therefore, we examine the number of citizen complaints that have resulted in enforcement actions; then we examine these complaints again, by region and over time from 1986 to 1989. Because some states are likely to have higher levels of complaints simply because they have higher levels of mining activity, all figures have been standardized by the number of inspectable units. The results appear in Table 11.3.

The results shown in Table 11.3 indicate that there is, again, a difference between the eastern and midwestern/western regions. In 1986, the incidence of complaints was almost twice as high in the East as in the other two regions. The West and Midwest had similar levels of complaints. From 1986 to 1989 the level of complaints fell in all regions, suggesting that either the quality of state enforcement or mining practices had improved.

In 1989 the greatest reductions in the number of citizen complaints was in the East, although it still had a greater number of citizen complaints because it initially had so many more than the other regions. Moreover, the difference between the East and the other two regions had declined somewhat by 1989. There is little reason to believe that the decline in complaints resulted from reduced activity by environmental groups, who usually file most of the complaints. If anything, the activity of those groups would have increased in the 1980s in response to the perceived lack of environmental concern by the Reagan administration.

A second means of evaluating the effectiveness of state enforcement is to examine federal oversight of state enforcement actions. When the federal government grants primacy to a state—gives the state the right to operate its own enforcement program—it does not relinquish its responsibility to implement the SMCRA. Federal oversight activities are conducted by OSM, which inspects mining operations in the states. When OSM identifies an improper mining activity, it issues a 10-day notice, informing the state agency it has detected a violation. The state has 10 days in which to issue a NOV or to show cause for not doing so. The greater the number of 10-day notices, therefore, the greater the number of improper mining practices that state has failed to identify.

The next analysis examines, by region, the number of federal 10-day notices issued. It would, however, be misleading and difficult to make comparisons among states and regions using only these numbers. The number of federal 10-day notices must be adjusted in two ways. First, because the number of OSM inspections differs substantially from state to state (more so than the number of state enforcement inspections mandated by OSM), those states with more inspections should receive more 10-day notices. The number of 10-day notices must, therefore, be standardized by the number of federal oversight inspections conducted in the state. Second, as was done for complaints, the data must be adjusted for the number of inspectable units (mines) in each state. The measure we used to examine federal oversight activity in each state, therefore, reflects the yearly number of 10-day notices that would be issued (at the current rate they are being generated) if each mine received one OSM inspection. The higher this number, the more infractions OSM is locating and the worse the quality of state enforcement. Again, federal enforcement data is examined over time to identify how increased federal commitment in the later 1980s affected the quality of state performance. The results appear in Table 11.4.

The data in Table 11.4 indicate that federal oversight activity has uncovered substantially higher levels of mining infractions in the East than

Table 11.4
Federal Oversight Score (By Mining Region)

Region	Oversight Score 1984-85 Average	Oversight Score 1988-89 Average
East	974.93	282.21
Midwest	140.17	44.07
West	12.34	18.18

Source: OSMRE Annual Reports.

All data adjusted for the number of federal oversight inspections and inspectable units.

in the other mining regions, particularly in the early 1980s. This suggests that eastern state inspectors missed large numbers of SMCRA infractions that federal enforcement agents detected. The best region was the West, which received a relatively low score. By 1989–90, the East, especially, made substantial progress in improving the quality of its enforcement. The oversight score in that region was one-third of its 1984–85 level. The Midwest also improved its enforcement by 1988–89; that region's score was about one-third its previous level. In the West, the overall score actually increased. The increase was small, and the overall score was consistently low, as would be expected of an area dominated by larger, corporate-owned mines that have the resources to comply with the provisions of the SMCRA (Harris 1985). Table 11.4 suggests that both the East and the Midwest responded to federal pressures. In spite of the improvement, however, eastern enforcement still lagged behind that in the other regions. The East was, apparently, still able to blunt some of the federal top-down pressures for better state enforcement.

CONCLUSION

This research indicates that both the federal government and state governments were able to influence implementation of the surface mining law. The federal government demonstrated its ability to influence state enforcement in the later 1980s. By the end of the decade, implementing officials had a greater commitment to insuring that national regulatory goals were achieved. OSM was able to push the East to increased levels of enforcement until it closely matched the performance of the West and Midwest.

These results also suggest that, following the Reagan administration's efforts to devolve greater regulatory authority to the states, the federal government was able to regain some control over implementation (Hedge, Scicchitano, and Metz 1988). This finding should reassure those who feared that devolution of regulatory authority to the states would preclude subsequent federal control. As we have seen, by the end of the 1980s, federal authorities were able to use available enforcement tools, most notably legislative requirements and oversight, to redirect the regulatory authority of eastern states.

It is similarly clear that federal actions do not tell the entire story of SMCRA enforcement. Substantial variations exist in levels of enforcement among the states (Scicchitano, Hedge, and Metz 1989). In the later 1980s, when federal officials had renewed vigor to achieve national regulatory goals, the quality of eastern enforcement still did not equal that of the other regions. Our several different measures of the quality and potential effectiveness of SMCRA enforcement all demonstrated that the East still lagged behind the West and the Midwest. This discussion of the quality of state enforcement also demonstrates the difficulty federal officials face in overseeing state programs. While federal officials can monitor the quantity of state enforcement relatively easily, it is more difficult to insure the quality of regulation.

Implementation literature can be valuable in explaining state SMCRA enforcement. The concept of the commitment of implementing officials, suggested by Mazmanian and Sabatier (1983), Bullock and Lamb (1984), and others, provides a useful perspective for explaining the actions of the federal government. Similarly, the Mazmanian and Sabatier (1983) notions of the technical difficulty of the task and the amount of behavioral change required provide useful insight into state SMCRA enforcement actions, particularly in the eastern mining region.

In the end, the complexity involved in implementing partial-preemption programs continues to amaze. We must be concerned with the influence of the federal government—an influence that changes over time—as well as with the actions of states. Clearly, we have much to learn about the ability and limits of each to control SMCRA implementation. Yet amid the complexity and confusion, we are beckoned to inquire further. Our nation's ability to resolve some its most pervasive environmental problems depends on a balanced and effective federal-state partnership—one that can successfully implement the surface mining law and other partial-preemption programs.

REFERENCES

Advisory Commission on Intergovernmental Relations. 1984. *Regulatory Federalism: Policy, Process, Impact, and Reform.* Washington, D.C.: U.S. Government Printing Office.

Bullock, Charles, III, and Charles M. Lamb. 1984. *Implementation of Civil Rights Policy.* Belmont, Calif.: Wadsworth.

Eads, George M., and Michael Fix. 1984. *The Reagan Regulatory Strategy.* Washington, D.C.: Urban Institute.

Harris, Richard A. 1985. *Coal Firms and the New Social Regulation.* Durham, N.C.: Duke University Press.

Hedge, David M., and Sabba Jallow. 1990. "The Federal Context of Regulation." *Social Science Quarterly* 71: 786–80.

Hedge, David M., Michael J. Scicchitano, and Patricia Metz. 1988. "State Regulatory Vigor in an Era of Deregulation." Paper presented at the 1988 meeting of the American Political Science Association, September 1–4, Washington, D.C.

Lilley, W., and J. Miller. 1977. "The New Social Regulation." *The Public Interest* 47: 49–61.

Mazmanian, Daniel A., and Paul Sabatier. 1983. *Implementation and Public Policy.* Glenview, Ill.: Scott, Foresman.

Menzel, Donald C. 1981. "Implementation of the Federal Surface Mining Control and Reclamation Act of 1977." *Public Administration Review* 41(2): 212–19.

———. 1983. "Redirecting the Implementation of a Law: The Reagan Administration and Coal Surface Mining Regulation." *Public Administration Review* 43: 411–20.

Nathan, Richard P., and Fred C. Doolittle. 1987. *Reagan and the States.* Princeton, N.J.: Princeton University Press.

———. 1984. "The Untold Story of Reagan's New Federalism." *The Public Interest* 77: 96–105.

Rosenbaum, Walter A. 1987. *Energy Politics and Public Policy.* Washington, D.C.: Congressional Quarterly.

Rowland, C. K., and Roger Marz. 1982. "Gresham's Law: The Regulatory Analogy." *Policy Studies Review* 1(3): 572–80.

Scicchitano, Michael J., David M. Hedge, and Patricia Metz. 1989 (Autumn). "The States and Deregulation: The Case of Surface Mining." *Policy Studies Review* 9: 120–31.

Shover, Neal, Donald A. Clelland, and John Lynxwiler. 1986. *Enforcement or Negotiation: Constructing a Regulatory Bureaucracy.* Albany, N.Y.: SUNY Albany Press.

Thompson, Frank J., and Michael J. Scicchitano. 1985. "State Enforcement of Federal Regulatory Policy: The Lessons of OSHA." *Policy Studies Journal* 13(3): 591–98.

U.S. General Accounting Office. 1986. "Surface Mining: Interior Department and States Could Improve Inspection Programs." Washington, D.C.: U.S. Government Printing Office.

Van Meter, Donald S., and Carl Van Horn. 1975. "The Policy Implementation Process: A Conceptual Framework." *Administration and Society* 6: 445–88.

Vestal, Theodore M. 1988 (Winter). "Federal Administration of the Surface Mining Control and Reclamation Act of 1977 in Oklahoma." *Publius* 18: 45–60.

Wassenberg, Pinky S. 1986. "Implementation of Intergovernmental Regulatory Programs: A Cost-Benefit Perspective." In J. Edwin Benton and David R. Morgan (eds.), *Intergovernmental Relations and Public Policy.* Westport, Conn.: Greenwood Press.

Welborn, David M. 1988. "Conjoint Federalism and Environmental Regulation in the United States." *Publius* 18: 27–43.

12 In the Eye of the Beholder: State and Federal Perceptions about the Surface Mining Control and Reclamation Act Enforcement

Denise Scheberle

INTRODUCTION

Like many environmental laws of the 1970s, the Surface Mining Reclamation and Control Act of 1977 (SMCRA) includes provisions for partial preemption, allowing federal and state governments to share regulatory responsibilities. A partial preemption approach gives states the option of assuming primary responsibility for running their regulatory programs (referred to as primacy). Once a state receives primacy, the federal partner adopts an oversight role, allowing the state to take the lead in regulatory enforcement. The utility of a partial preemption regulatory design for the implementation of environmental laws has been criticized (Squillace 1985; Crotty 1987; Davis and Lester 1989; Gormley 1987).

One reason that partial preemption prompts criticism is that it depends upon two potentially conflicting factors: (1) the need for positive intergovernmental working relationships and (2) the need for effective federal oversight of the primacy states.

Squillace (1985) argued that an "undeniable strain" has developed between state agencies and the Office of Surface Mining (OSM), and this undisguised tension can only result in continued confrontation and backsliding from the congressional mandate of SMCRA. Harris (1989) and Miller (1989) echo the difficulties of achieving cooperative federalism under SMCRA and suggest that federal regulators must walk the difficult line between being an overseer and not adopting inflexible implementation positions.

Earlier studies of intergovernmental relationships between the Office of Surface Mining, the federal agency charged with oversight under SMCRA, and state regulatory agencies suggest little correspondence in attitudes between federal inspectors and their state counterparts (Hedge et al. 1988; Shover et al. 1986).

Hedge, Menzel, and Williams (1988) found a difference in the role orientations of OSM inspectors and state inspectors. In their approach to enforcement, federal inspectors tended to take a "going by the book" legalistic approach; state inspectors tended to take a more accommodating approach. Shover, Clelland, and Lynxwiler (1986) also noted that OSM inspectors tended to prefer enforced compliance while state inspectors were more open to negotiation. This attitudinal difference hampered intergovernmental cooperation in enforcement efforts within coal states.

These studies indicate that important perceptual differences exist between federal inspectors, who have an oversight role, and state inspectors, who are charged with on-the-ground enforcement efforts under partial preemption. However, the studies by Hedge, Menzel, and Williams (1988) and Shover, Clelland, and Lynxwiler (1986) were conducted during the early years of SMCRA implementation. In order to more fully assess the utility of a partial preemption approach, it is important to observe whether perceptual differences between OSM inspectors and state inspectors have persisted as SMCRA has matured.

THE STUDY

This study explores the effectiveness of a partial preemption regulatory approach as perceived by federal and state inspectors after nearly a decade of state primacy and thirteen years after the passage of SMCRA. Specifically, it looks at the perceptions that OSM inspectors and state inspectors have about the success of the regulatory program, the intergovernmental working relationship, the oversight ability of the federal agency, and major obstacles (if any) that hinder the regulation of mining activities in coal states.

Several research questions guide the study. First, what is the nature of the intergovernmental working relationship? Is there a vertical alignment within a primacy state—something like "picket fence" federalism, where federal and state inspectors share similar concerns about the nature of the problem, have mutual respect for each other, and identify the same types of obstacles to achieving environmental goals (Wright 1988)? Or, conversely, do inspectors have a horizontal alignment? That is, are state inspectors close to regulated industry and to other state inspectors? Do they have competing goals other than those embodied in the regulatory policy (i.e., economic development)? Finally, are disharmony and distrust of the federal oversight agency evident?

If the vertical alignments posited by cooperative federalism in fact do exist in the minds of federal and state inspectors, they should agree on several important regulatory matters. First, they should see the performance of both the state and the OSM as roughly equal. If they perceive that one agency is vastly superior (or exceedingly inept) in comparison to the other one, the assumptions of partial preemption are violated (i.e., the state regulatory agency is an adequate surrogate for a federal program).

Second, both federal and state inspectors should reach some degree of consensus about the nature of the environmental problem and the nature of the target group, the coal operators within that particular coal state. Similarly, inspectors should hold similar views about what comprises the most troublesome obstacles to effective SMCRA implementation.

Finally, vertical alignment suggests a minimum of intergovernmental animosity between actors charged with street-level enforcement—federal and state inspectors. Inspectors should perceive a federal-state relationship that is mutually beneficial and reciprocal. If state inspectors perceive federal inspectors as overly aggressive in their oversight, or if federal inspectors see their state counterparts as ineffective, then vertical alignment becomes problematic.

On the other hand, indicators of horizontal alignment would include perceptions of a closer relationship between coal operators and state inspectors than between federal and state inspectors, perceptions of confrontation and lack of communication between federal and state regulatory agencies, and perceptions of general disdain among state inspectors for the performance of the federal oversight agency. The presence of these characteristics within the state agencies suggests a breakdown of the tenuous federal-state working relationship.

In order to test these research expectations, survey and elite interview data were gathered from state regulatory personnel in Colorado, Wyoming, and Virginia, and from federal inspectors in three corresponding OSM field offices. Questionnaires were delivered to all federal and state agency personnel involved in coal mining regulatory enforcement in the three states. Questionnaire return rates were 87 percent (74 out of 85 surveys) for state personnel and 87.5 percent (14 out of 16) for federal inspectors.[1] In addition to the survey research, a total of 57 interviews were conducted with administrators and inspectors from all three OSM field offices and with state administrators and inspectors from Wyoming, Virginia, and Colorado.

Table 12.1
Evaluations of State Agency Performance and OSM Performance

Evaluation	Percentage that Strongly Agree	
	State Inspectors (n = 74)	OSM Inspectors (n = 14)
State Evaluation		
State agency does a good job	80	36
State is concerned about environmental damage	77	7
State inspectors cite all violations	60	21
State agency is too lenient in regulating coal operators	8	57
State regulations are consistently enforced	37	0
State Agency is more interested in producing coal than protecting the environment	29	71
OSM Evaluation		
OSM does a good job	34	64
OSM is concerned about environmental damage	54	36
SOM is too lenient in regulating coal operators	3	64
OSM is more interested in coal production than environmental protection	5	43

FINDINGS

Perceptions about OSM and State Agency Performance

A comparison of responses to scaled survey questions about state agency performance and OSM performance indicates little consensus between federal inspectors and state inspectors (see Table 12.1). Not surprisingly, state inspectors feel much better about their state regulatory programs than about the OSM program, and OSM inspectors evaluate their agency more highly than do their state counterparts. While 80 percent of the state inspectors strongly agree that their state agency is doing a good job at regulating coal operators, only 36 percent of OSM inspectors do.

More importantly, federal and state inspectors alike show little alignment across a number of evaluation criteria for the state agency. Only 7 percent of the OSM inspectors feel that the state is very concerned about the environmental consequences of mining coal, while 77 percent of state inspectors do (93 percent of the Colorado inspectors, 75 percent of the Wyoming inspectors and 72 percent of the Virginia inspectors). Only 21 percent of the OSM staff feels that state inspectors regularly cite violations (compared to 60 percent of state inspectors). In regard to regulatory approach, 57 percent of the OSM inspectors see their state counterparts as too lenient (compared to 8 percent of state inspectors). In addition, none of the OSM inspectors feel that state inspectors apply regulations consistently to all coal operators (compared to 37 percent of the state inspectors).

Thus, it appears that there is little agreement between state and federal inspectors over state performance or state posture toward the coal industry. Federal inspectors tend to evaluate state performance more negatively than do their state counterparts.

When the OSM is evaluated, a similar pattern emerges. OSM inspectors tend to evaluate OSM performance more highly than do state inspectors (see Table 12.1). Only 34 percent of state inspectors believe that the OSM is doing a good job (compared to 64 percent of OSM inspectors). However, federal inspectors are more critical of the performance of the OSM than their state counterparts are of state agencies (Table 12.1). For example, while only 8 percent of state inspectors judge their state agency to be too lenient, a substantial 64 percent of the OSM inspectors judge the OSM to be too lenient. A paltry 3 percent of state inspectors believe that the OSM is too lenient.

Furthermore, state inspectors tend to see the OSM as more attentive to environmental goals than OSM inspectors do. While 95 percent of state inspectors believe the OSM is oriented more toward protecting the environment than producing coal, only 57 percent of the federal inspectors do. Only 36 percent of the OSM inspectors believe that their agency is concerned about stopping environmental damage associated with coal mining while 54 percent of the state inspectors do.

While state inspectors tend to see their regulatory agency as somewhere in the middle, between favoring environmental protection or industrial development, they see state legislators as clearly pro-industry. Among state inspectors, about 30 percent believe that the state agency favors industrial development, while 64 percent believe that state legislators favor industrial development over environmental protection. The OSM, on the other hand, sees a much stronger pro-industry position for both the state agencies and the state legislature. Of the OSM inspectors, 71 percent

Table 12.2
Perceptions about the Coal Operators and Environmental Problems

Perception	Percentage that Strongly Agree	
	State Inspectors (n = 74)	OSM Inspectors (n = 14)
Coal operators are concerned about the environment	19	14
Reclamation efforts are better	77	50
Coal mining is a serious environmental threat	20	50

believe state agencies favor coal production over environmental protection, and 86 percent believe that state legislators favor coal production.

In summary, OSM inspectors do not share the positive assessment that state inspectors have of the state agency's regulatory performance and dedication to SMCRA requirements. However, OSM inspectors do not provide an overwhelming endorsement of their own agency either, seeing it as too forgiving of coal operator violations and too lenient in its oversight of state regulatory agencies. State inspectors, on the other hand, see the OSM as more environmentally sensitive than their own agencies. Few state inspectors, however, believe that the OSM is doing a good job in its oversight role.

Perceptions about Environmental Problems and Coal Operators

A second indicator of vertical alignment between federal and state inspectors is the extent to which they share similar perceptions about the nature and seriousness of environmental problems and about the goodwill efforts of the target group—in this case, the coal operators. Here, too, are indications of differing perceptions, as the data in Table 12.2 shows. Generally, OSM inspectors see less improvement in reclamation efforts than state inspectors do. Only half agree that reclamation efforts have improved in the last few years, compared to 77 percent of the state inspectors. While 50 percent of the federal inspectors view coal mining as a serious environmental threat, only 20 percent of state inspectors do.

The only point of consensus between federal inspectors and state inspectors is in their evaluation of the coal operators. Less than 20 percent of either federal or state inspectors characterize coal operators as strongly

Table 12.3
Perceptions of the Relationship between OSM and State Agency

Perception	Percentage that Strongly Agree	
	State Inspectors (n = 74)	OSM Inspectors (n = 14)
OSM gives the state too much discretion	5	57
Without OSM oversight, state would not enforce SMCRA	28	86
OSM should be more involved in enforcing regulations	7	43
OSM-state relationship is confrontational rather than friendly	68	21
State is less suited than the OSM to implement regulatory program	4	43
Communication with the OSM has gotten worse	57	*
OSM oversight efforts have improved	23	*

*Questions given only to state inspectors.

environmentally concerned. But the federal-state perceptual dichotomy is again evident when coal operator motives are evaluated. Of those surveyed, 86 percent of OSM inspectors believe that coal operators are very influential in state regulatory processes; only 39 percent of state inspectors do. OSM inspectors are three times more likely than are state inspectors to suspect that coal operators intentionally violate SMCRA regulations (47 percent to 13 percent).

Thus, OSM field staff see both the nature and the extent of environmental problems associated with coal mining as more serious than do their state counterparts. They also feel that less progress in reclamation has been made than do state inspectors. Both see the need to monitor coal operator behavior, but OSM inspectors are more likely to be suspicious of coal operator intentions and of their ability to influence the state's regulatory program.

Perceptions about the Success of Cooperative Federalism

Probably the most important indicator of vertical alignment between state and federal inspectors is the way they view their intergovernmental relationship. Responses to survey questions that measure perceptions about federal-state working relationships clearly indicate a less-than-harmonious atmosphere. The data presented in Table 12.3 show that state inspectors believe communication between their agency and the OSM has gotten worse over the last few years; furthermore, only 23 percent of the state inspectors agree that OSM's oversight has improved.

State and OSM inspectors disagree about the extent of federal involvement in state regulatory programs. While 89 percent of state inspectors believe that OSM should be less involved in enforcing SMCRA regulations within their state, 43 percent of OSM inspectors believe they need to be more involved in state enforcement. State inspectors also strongly disagree with the suggestion that the state is less suited to regulate coal mining than is the federal government (93 percent of the Colorado inspectors, 83 percent of the Wyoming inspectors, and 85 percent of the Virginia inspectors). More than 40 percent of the OSM inspectors, on the other hand, strongly agree that the state is less suited to implement regulatory policy than is the federal government.

When asked whether the nature of the OSM-state relationship was confrontational, 68 percent of the state inspectors (72 percent of the Virginia inspectors, 73 percent of the Colorado inspectors, and 55 percent of the Wyoming inspectors) strongly agree. In contrast, only 21 percent of the OSM inspectors agree. Finally, when asked whether OSM gives the state too much discretion, state inspectors answered in concert: only 5 percent agreed (100 percent of the Colorado inspectors, 84 percent of the Wyoming inspectors, and 98 percent of the Virginia inspectors strongly disagree). In contrast, 57 percent of OSM inspectors believe that primacy states have too much discretion in running their regulatory programs.

This evaluation of the breakdown of the federal-state regulatory relationship under primacy is also evident when inspectors were asked to identify roadblocks to SMCRA implementation. Table 12.4 lists the five implementation roadblocks most frequently identified by federal and state inspectors.[2] Both federal and state inspectors identify ongoing disagreements between agencies as a significant obstacle to effective implementation.

The contentious federal-state relationship is the implementation obstacle most frequently identified by state inspectors. This intergovernmental confrontation is more important to a majority of state inspectors than is any other political, economic, or technical variable. To state inspectors, the source of federal-state friction is OSM's belief in enforced compliance, an approach that contrasts with the more accommodative regulatory approach preferred by state inspectors. Too often, in the eyes of state inspectors, the OSM regulatory approach translated into oversight that

Table 12.4
Roadblocks to Implementation

Ranked by state inspectors:

1. Disagreements between state and OSM personnel
2. Lack of policy guidance from state administrators
3. Frequent changes in federal rules and regulations
4. Lack of support by state legislators
5. Resistance by coal firms

Ranked by OSM inspectors:

1. Lack of policy guidance from state administrators
2. Disagreements between state and OSM personnel
3. Too little discretion given to state inspectors*
4. Lack of policy guidance from OSM*
5. Conflicting goals among various state departments

*Tied in ranking results.

concentrated on "paper" compliance rather than on substantial regulatory results. As two Colorado inspectors noted:

> Adversarial roles should be down played. OSM should be a friendly overseer rather than a "big brother" ready to beat us with a stick. . . . OSM Field Offices should provide less policing of our state program and more technical assistance. Unfortunately, OSM headquarters likes the "big brother" approach. The OSM seems more interested in bean-counting than on the big environmental picture.
>
> I would like to prevent OSM inspectors from citing frivolous violations. They spend more time issuing enforcement actions [such as] requiring a 16 inch rather than an 18 inch culvert than on significant violations that may affect water quality or species diversity. (From written survey response)

Virginia inspectors are especially sensitive to what they feel is an overbearing federal presence, as indicated by the following comments:

> OSM inspectors are only interested in making state inspectors look bad. The OSM and the state agency should be in existence for the same purpose, instead we are in constant combat.
>
> OSM is too concerned about numbers which do not accurately reflect the job being done in Virginia. Paperwork is more important to them (the OSM) than what's being done in the field. (Interviews with personnel from Virginia Department of Mines, Minerals and Energy, June 1990)

Other responses indicate perceptions among state inspectors of vindictiveness in OSM oversight. These state inspectors believe that the OSM is not just "going by the book"; it is trying to undermine the state agency, as illustrated by the following comments:

> The state has responsible people running a program that has been approved by the federal government. The effectiveness of our program will neither be diminished nor improved by the constant attempts of the OSM to *discredit* our program. (From written survey response; emphasis added)

> Life would be a lot simpler and our efforts more effective if the OSM Field Office would stop trying to make us look bad. (Interview with state inspector, June 1990)

Other implementation obstacles cited by state inspectors include the lack of policy guidance from state administrators, lack of political support from state legislators, and frequent changes in federal rules and regulations. As with OSM-state conflicts, these political variables were perceived to be more important than the resistance of coal firms or state resources such as adequate staffing and funding levels.

State inspectors who received little policy guidance from agency administrators reasoned that the lack was due to intrastate political pressure on the agency. In some instances, intrastate political pressure distorted the genuine desire of state inspectors to be reasonable in enforcement (i.e., to distinguish serious violations from less serious violations, and to overlook minor infractions for the sake of fostering a better working relationship with coal operators) (Bardach and Kagan 1982). In other cases, even though failure to cite all *observed* violations is a direct violation of SMCRA, pressures emanating from the top of state government (state legislators or the state executive), and percolating down from the top of the regulatory agency, made state inspectors reluctant to issue citations for fear they would be vacated by state administrators, as illustrated by this comment:

> The absence of administrative support all too often results in administrators' changing inspector's decisions. Sometimes, decisions [about enforcement] are so political that administrators worry about how coal operators, politicians or political appointees will view the decision—that generates uncertainty and inconsistency in the field. (From written survey response)

One state inspector noted that he "walked a tightrope" between complying with OSM requirements and satisfying the state's "overriding desire

Table 12.5
Recommendation for Improving Regulatory Performance

State Agency Improvements*
<u>OSM recommendations</u>
1. Remove political pressure from state administrators; reduce politics at upper management levels of state agency
2. Do away with compromise; return to the letter of the law; exact application of regulations
3. Cite all violations observed; increase enforcement efforts
<u>State recommendations</u>
1. Reduce or remove OSM oversight
2. Remove political pressure from state administrators
3. Increase regulatory flexibility; regulations should be individually applied to sites and/or states
4. Establish fixed civil penalties; increase bond amounts; accept only sound bond instruments; increase civil penalties
OSM Field Office Improvements*
<u>OSM recommendations</u>
1. Increase cooperation between state and OSM personnel
2. Remove politics from OSM and let OSM do its job impartially
3. Increase oversight role of OSM
4. OSM personnel should follow OSM policies
<u>State recommendations</u>
1. Prevent OSM inspectors from citing frivolous violations; OSM overly focused on paperwork and statistical compliance; too rigid in regulatory applications
2. Eliminate or reduce the oversight role of the OSM
3. Increase cooperation between state and OSM personnel

*Ranked in order of response frequency.

for economic development." He added this statement about his enforcement activities:

Cite all violations? Why should I? Most NOVs (Notices of Violation) that I write will just be vacated by management anyway. We [state inspectors] can't really make enforcement decisions on our own, for fear of being shot out of the saddle by upper management. (From elite interview, July 1990)

Thus, intrastate political pressures cause Colorado, Wyoming, and Virginia inspectors to carefully consider their enforcement decisions. State inspectors identify it as the second major obstacle to implementation.

Not surprisingly, federal and state inspectors offer different recommendations for improving regulatory performance under SMCRA (see Table 12.5). OSM inspectors most frequently suggest eliminating the political pressures that state administrators place on state inspection staff. Federal inspectors often believe their state counterparts are under considerable pressure not to write tickets or otherwise unduly hamper coal operators. One OSM inspector put it this way:

> What the Department of Environmental Quality is required to do as far as regulations (under SMCRA) and what it is told to do by state administrators are diametrically opposed. The end result is [the achievement of] short-term political accommodation, not technically-sound reclamation in the real world in Wyoming. . . . The present emphasis on cosmetic enforcement will cause Wyoming to be known as a "National Sacrifice Area" in the year 2015. (From survey response)

In addition to removing intraagency political pressure on state inspectors, OSM inspectors argue for a return to a more enforced regulatory approach—eliminating bargaining with coal operators, restricting flexibility in citing violations, and generally returning to the letter of the law. Federal inspectors wish to combat state regulatory malaise with firm adherence to SMCRA regulations. As argued by one OSM field inspector, "the OSM is too timid about being sure that the [state] program is in full compliance. The OSM should assure that full compliance is achieved—nothing more, nothing less" (Conversation with OSM field inspector, June 1990).

State inspectors, however, most frequently recommend that the role of the OSM be either reduced or completely eliminated, as indicated by these comments (see also Table 12.5):

> Take the OSM completely out of Colorado. I believe the state spends 25 percent of its time sparring with the OSM. Eliminate the OSM. The OSM is a classic example of a federal agency set up to "set up" state agencies. If the OSM finds itself moving toward extinction, it will dream up unnecessary requirements just to keep its people employed. (Conversations with state inspectors, August 1990)

In sum, it appears that neither federal field office inspectors nor state inspectors from eastern or western states have positive perceptions of the intergovernmental working relationship. Their perceptions are so negative, in fact, that both sets of inspectors see the federal-state relationship as the preeminent constraint on SMCRA implementation. State inspectors feel, for the most part, that SMCRA implementation would be better if the OSM would reduce or refocus their oversight; OSM inspectors feel that state inspectors should steadfastly adhere to the regulatory dictates of SMCRA.

CONCLUSIONS

The survey results indicate that the intergovernmental harmony suggested by cooperative or "picket fence" federalism does not describe the relationship between inspectors from three OSM field offices and three coal states. State and federal inspectors exhibit markedly different attitudes in all three areas of possible vertical alignment: the performance of the state and federal agencies, the nature of the target group and extent of the environmental problem, and the quality of the federal-state working relationship.

The most universal perceptions among inspectors from Wyoming, Colorado, and Virginia are their negative evaluations of OSM oversight and OSM intervention into state regulatory programs. Thus, the data support the observations of Squillace (1985) and others that there is little "partnership" in the federal-state regulatory alignment. Also, as indicated by Hedge, Menzel, and Williams (1988) and Shover, Clelland, and Lynxwiler (1986), state and federal inspectors have distinctly different approaches to fulfilling SMCRA regulatory requirements. Furthermore, while a certain amount of conflict might be expected in any oversight relationship, the results of this study indicate that the frustration gap between federal and state inspectors in this policy area is far wider than one might expect. This conflict is evident in both eastern and western coal mining regions and has remained, even though primacy under SMCRA is now almost a decade old. Contentious intergovernmental relationships have persisted through the Reagan years, when decentralization and New Federalism were the rallying calls of federal administrators (Menzel 1983).

Finally, the research findings indicate an intraorganizational dissatisfaction and lack of consensus about agency goals and the organizational mission between OSM headquarters and the three field offices. A substantial percentage of OSM field inspectors seem to be in the uneasy position of favoring a regulatory approach that is not supported by their administrators or by the primacy states. Thus, they are frustrated from above and below.[3] In turn, confusion within the oversight agency creates continuing administrative confusion for primacy states.

One OSM inspector chastised his own agency, saying, "The OSM has been in a state of turmoil and confusion since its conception. The staff and management use the agency to manage by vindictiveness rather than by objectives. All one needs to see is the moves made by OSM personnel and it becomes vividly apparent" (Interview, Virginia OSM Field Office, June 1990).

Given the results of this research, for surface mining regulatory implementation at least, several questions are worth reflection. Why is this relationship so hostile, and why has this hostility persisted over nearly a decade without much abatement? Further, is the degree of intergovernmental conflict the typical dynamic of federal-state relationships, or is it peculiar to SMCRA implementation? Finally, what are the consequences of this perceptual gap between federal and state implementors for partial-preemption regulatory schemes?

The continued and pronounced lack of vertical alignment between federal and state inspectors may be the result of several interacting factors unique to SMCRA. First, OSM field office personnel clearly embrace the strict enforced compliance style of the early OSM, even though subsequent OSM political leaders have called for increased delegation to state regulatory agencies. The adherence to enforced compliance regulatory styles may be due to several factors. Without administrative support, OSM inspectors may prefer to walk the straight and narrow regulatory path that is applauded by environmental and citizen groups. Environmental groups have typically focused their efforts at enforcing SMCRA through federal oversight. Congressional critics have regularly identified a lack of vigorous OSM oversight of state primacy programs (U.S. Congress 1987).

State inspectors, on the other hand, most definitely know the rules of the game for winning (or losing) the support of their agency administrators. From the beginning, state administrators have supported an accommodative regulatory approach toward coal operators. Unlike the OSM, state agencies have not changed their regulatory posture. Thus, where federal inspectors have had to face intraorganizational conflict about regulatory attitudes, state inspectors have faced consistent regulatory attitudes.

Whether state inspectors totally embrace their state's regulatory philosophy is, however, another matter. As mentioned earlier, many state inspectors share a common perception: that there is unilateral political pressure on state enforcement agencies to accommodate coal companies. As stated by one state inspector, "This agency bends with the political winds, and the winds almost always come from the coal companies" (Interview, August 1990).

Although state and federal inspectors apparently prefer distinctly different enforcement styles, there is a nuance to the black-and-white perceptions of state inspectors. That is, while nearly all state inspectors believe that the OSM has gone too far in its oversight role, they also experience intraagency pressure to bias their enforcement styles toward accommodation, above and beyond what they otherwise would embrace. Therefore, despite obvious frustration with the OSM bean-counting oversight approach, many state inspectors believe they need the federal presence as a counterweight to internal agency pressures. The use of the OSM as an impetus for enforcing SMCRA was revealed during one inspection when the state inspector told the coal operator, "I'd overlook this [violation], but you know the OSM guys won't." This inspector said that he frequently uses the OSM as a "heavy" when he issues citations (Interview, June 1990).

Another factor that has worked to increase intergovernmental conflict has been the massive rewriting of OSM regulations. Early in the Reagan administration, OSM Director James Harris criticized the agency's overly burdensome and heavy-handed approach to primacy states (U.S. Department of the Interior 1984). By 1983, 91 percent of OSM regulations had been rewritten (U.S. Department of the Interior 1984). Thus, not only did state inspectors have to contend with OSM's "going by the book" attitude in the field, it was hard for both state inspectors and OSM personnel to know exactly what *was* in the "book" of federal regulations.

Finally, characteristics intrinsic to SMCRA tend to exacerbate intergovernmental conflict under cooperative federalism. For example, SMCRA targets a specific industry and mandates specific and uniform standards of performance for each mine operator. States, therefore, do not enjoy the same level of flexibility under SMCRA as they do under other environmental programs. For example, under the Clean Air Act, states can develop their own state implementation plans to meet national ambient air quality standards, rather than enforce one uniform standard on each emitting source (Squillace 1985, 696).

What is the prognosis for cooperative federalism under SMCRA? One can speculate that intergovernmental cooperation will remain elusive. Cooperative federalism under SMCRA clearly does not represent primacy's finest hour—nor is it ever likely to, given the current complex of regulatory enforcement patterns. It is not unreasonable to expect, therefore, that the lack of cooperation among federal and state participants will continue to hamper SMCRA's implementation process.

It is beyond the scope of this study to fully explore remedies for the strained federal-state relationship, but a few concluding suggestions can be made. One obvious solution would be to eliminate state regulatory activities and return to direct federal enforcement of SMCRA in coal states. This would, however, require Congress to change the act, which is unlikely given congressional preference for primacy arrangements. In addition, intraorganizational problems within OSM, as noted by OSM participants in this study, would not bode well for such a dramatic increase in federal activities. It is unclear whether a stricter federal command-and-control regulatory posture would ultimately result in improved compliance with SMCRA provisions. And it is worth repeating that a significant portion of state agency personnel do not wish to abandon the federal-state partnership: they wish to change its character.

Both state and federal inspectors see themselves as environmentalists, and both are concerned about environmental quality; however, they disagree vehemently about how to achieve it. Thus, resentment among state field staff might decrease if OSM field staff communicated their common desire to achieve qualitative on-the-ground environmental results, rather than merely quantitatively monitoring state enforcement efforts.

Second, improved cooperation and support within the OSM itself might lead to improved intergovernmental cooperation. If OSM field staff and OSM headquarters staff would work to reestablish intraagency trust, OSM field personnel would not believe they had been abandoned.

Finally, state officials should work to alleviate political pressures on state inspectors by supporting field enforcement decisions. In turn, federal oversight inspections would be less tense; then, perhaps, the intergovernmental working relationship postulated by cooperative federalism could begin.

NOTES

1. The return rates for individual state agencies are as follows: the Land Quality Division, Department of Environmental Quality in Wyoming, 100 percent (20 surveys returned out of 20 mailed); Department of Mines, Minerals and Energy in Virginia, 87 percent (40 out of 46); Division of Mined Land Reclamation in Colorado, 78 percent (14 out of 18). The three OSM Field Offices are located at Big Stone Gap, Virginia; Casper, Wyoming; and Albuquerque, New Mexico (which oversees the Colorado regulatory program).

2. The survey question listed 16 possible obstacles to implementation of SMCRA. Respondents were asked to identify the three obstacles, or roadblocks, they felt were most important. The question also allowed for open-ended responses.

3. Although not part of the present discussion, it is worth noting that nearly all of the environmental groups surveyed as part of the larger study supported the enforced compliance posture of the OSM field inspectors and urged the OSM headquarters to, as one environmental group member put it, "get some backbone."

REFERENCES

Bardach, Eugene, and Robert A. Kagan. 1982. *Going by the Book: The Problem of Regulatory Unreasonableness*. Philadelphia: Temple University Press.

Crotty, Patricia. 1987. "The New Federalism Game: Primacy Implementation of Environmental Policy." *Publius* 17: 53–67.

Davis, Charles E., and James P. Lester. 1989. "Federalism and Environmental Policy." In James P. Lester (ed.), *Environmental Politics and Policy: Theories and Evidence*. Durham, N.C.: Duke University Press.

Goggin, Malcolm L., Ann O'M Bowman, James P. Lester, and Laurence O'Toole. 1990. *Implementation Theory and Practice: Toward a Third Generation*. Glenview, Ill.: Scott, Foresman and Co.

Gormley, William. 1987. "Intergovernmental Conflict on Environmental Policy: the Attitudinal Connection." *Western Political Quarterly* 40: 285–303.

Harris, Richard. 1989. "Federal-State Relations in the Implementation of Surface Mining Policy." *Policy Studies Review* 9: 69–78.

Hedge, David M., and Donald C. Menzel. 1985. "Loosening the Regulatory Ratchet: A Grass-Roots View of Environmental Deregulation." *Policy Studies Journal* 13: 599–606.

Hedge, David M., Donald C. Menzel, and George H. Williams. 1988. "Regulatory Attitudes and Behavior: The Case of Surface Mining Regulation." *Western Political Quarterly* 41: 323–40.

Menzel, Donald C. 1983. "Redirecting the Implementation of a Law: The Reagan Administration and Coal Surface Mining Regulation." *Public Administration Review* 43: 411–20.

Miller, Richard. 1989. "Implementing a Program of Cooperative Federalism in Surface Mining Policy." *Policy Studies Review* 9: 79–87.

Rosenbaum, Walter. 1983. *Environmental Politics and Policy*. Washington, D.C.: Congressional Quarterly Press.

Shover, Neal, Donald A. Clelland, and John Lynxwiler. 1986. *Enforcement or Negotiation: Constructing a Regulatory Bureaucracy*. Albany: SUNY Press.

Squillace, Mark. 1985. "Cooperative Federalism Under the Surface Mining Control and Reclamation Act: Is This Any Way to Run a Government?" *West Virginia Law Review* 87: 687–708.

U.S. Congress House Committee on Interior and Insular Affairs. 1987. *Tenth Anniversary of the Surface Mining Reclamation and Control Act of 1977*. Oversight Hearings Before the House Subcommittee on Energy and the Environment, 100–126. Washington, D.C.: U.S. Government Printing Office.

U.S. Department of the Interior. 1984. *OSM Annual Report for Fiscal Year 1983*. Washington, D.C.: U.S. Government Printing Office.

Wright, Deil S. 1988. *Understanding Intergovernmental Relations*. 3d ed. Pacific Grove, Calif.: Brooks Cole Publishing Co.

13 The Surface Mining Control and Reclamation Act: Changes in State Capabilities

Kathryn M. Mutz and Ronald W. Daniels

INTRODUCTION

The Surface Mining Control and Reclamation Act (SMCRA) invites coal-producing states to participate in a nationwide program to "protect society and the environment from the adverse effects of surface coal mining operations" (Public Law 95–87, section 102[a] 30 U.S.C. 1201).

Of the 27 states that currently produce coal (Energy Information Administration 1989), 24 states currently have or have had primacy under SMCRA. Most of the primacy states have operated approved programs for periods of time ranging from 8 years to 12 years, with an overall average of about 10 years. Of those 27 states, all but two (Alaska and Louisiana) had some form of coal regulatory program before SMCRA was enacted (see Table 13.1). States that received primacy either modified their existing programs or developed new ones that were as stringent as the federal program. Tennessee had primacy but relinquished it and repealed its state enabling act.

The ability of the states to control the impacts of surface mining and the above-ground impacts of underground coal mining depends on having appropriate laws and regulations, adequate staff, and the political will for enforcement. How has that ability changed with the implementation of SMCRA? Because it is difficult to evaluate political will, which involves not only publicly espoused agency policy but also legislative funding and informal agency operations, this chapter will concentrate on changes and perceived changes in laws and regulations and in state program staffs.

Table 13.1
State Coal Programs before SMCRA[a]

Region and State	Years of Coal Mine Regulation	Staff Size	Mines		Production	
			Number	Staff	Total[b]	Mine
EASTERN STATES		553[d]	5,896[d]	10.7[d]	550,325[d]	93[d]
Appalachian Coal Region		481	5,706	11.9	444,563	78
Alabama	21	10	236	23.6	22,644	96
Kentucky	37	200	2,401	12.0	143,613	60
Maryland	36	9	69	7.7	2,606	38
Ohio	43	80	348	4.4	46,770	134
Pennsylvania	46	100	835	8.4	84,137	101
Virginia	25	32	745	23.3	35,510	48
West Virginia[c]	55	50	1,072	21.4	109,283	102
Average of Appalachian States	38	69	815	14.4	63,509	83
Interior Coal Region		72[d]	190[d]	2.6[d]	105,762[d]	557[d]
Arkansas	10	1	8	8.0	488	61
Illinois	29	12	58	4.8	59,537	1,027
Indiana	50	10	62	6.2	25,124	405
Iowa	50	2	10	5.0	622	62
Kansas	22	5	4	0.8	479	120
Louisiana	10	0	0	--	0	0
Missouri	19	4	13	3.3	5,638	434
Oklahoma	84	10	31	3.1	2,872	93
Texas	16	28	4	0.1	11,002	2,751
Average of Interior States	32	9[d]	24[d]	3.9[d]	13,220[d]	619[d]
Average of all Eastern States	35	37[d]	393[d]	8.8[d]	36,688[d]	369[d]
WESTERN STATES AND WESTERN COAL REGION		51[e]	97	1.9[e]	79,104	816
Alaska	8	0	1	--	766	766
Colorado*	18	10	33	3.3	8,219	249
Montana*	18	8	8	1.0	22,054	2,757
New Mexico*	18	7	5	0.7	8,785	1,757
North Dakota*	21	6	10	1.7	8,515	852
Utah*	16	3	20	6.7	6,961	348
Wyoming*	18	17	20	1.2	23,804	1,190
Average of Western States	17	9[e]	14	2.4[e]	11,301	1,131

*Member of Western Interstate Energy Board.
[a]Data on coal production and mines is for 1975; staff size is "before SMCRA."
[b]Thousands of short tons.
[c]Data were obtained from telephone interview rather than questionnaire.
[d]Does not include Louisiana.
[e]Does not include Alaska.

Information and opinions were solicited from administrators of coal regulatory programs throughout the country and from past and present agency personnel. Questionnaires were distributed to states and to tribes with coal resources. All but one state and three tribes responded to the questionnaire, which consisted of three pages that could be completed in less than two hours. Responses from twenty-two states that currently have primacy under SMCRA form the basis of the conclusions drawn from the survey. Data from states and tribes without primacy are not included. In the following text, *italic type* is used to indicate multiple-choice questionnaire responses.

Questionnaire respondents are not identified. However, references are made to eastern and western trends and to the Appalachian, interior, and western coal regions. The distinction was made between eastern and western coal states because of the perception that SMCRA was passed largely in response to mining abuses in Appalachia, thus causing an eastern law to be imposed on western states. This chapter describes the effect SMCRA has had on state regulatory programs. Whether this effect differs from East to West or is independent of geographic location will be discussed.

The 100th meridian has been chosen as the dividing line between western states and eastern states for purposes of this discussion. The eastern states include primacy states of the Appalachian and interior coal regions; the western states include primacy states of the western coal region (Energy Information Administration 1989). Table 13.1 provides pooled statistics for the eastern states, as well as separate data for the Appalachian and interior coal regions. While the pooled data for the eastern states display the same trends as the data for the Appalachian region, statistics for the interior region indicate many similarities to the western region.

In addition to information from the questionnaires, supplemented by interviews with agency personnel, state regulatory programs of two western states (Utah and Wyoming) are discussed in detail. Both states are politically conservative and tend to elect conservative Republicans to their state legislatures and congressional delegations, although both had Democratic governors in the early years of SMCRA. Wyoming's current governor is also a Democrat. Both Wyoming and Utah have small populations (Wyoming's 1987 population was 490,000; Utah's was 1,680,000), and both have vast tracts of unpopulated lands. Extraction of natural resources is a significant economic activity in each state, and each has a large proportion of federal lands and minerals. Because both states have large areas of federal lands, their coal programs are largely subsidized by federal grants. Both states have cooperative agreements with OSM whereby the state administers SMCRA on federal lands and OSM provides 100 percent of the cost of program administration for these federal lands (SMCRA section 704[c]).

Coal mining in these states differs, however. Utah, with a long history of small, underground mines, currently has 21 underground mines producing about 20 million short tons, while Wyoming, dominated by large, relatively recent surface mines, has four underground mines producing 1.6 million short tons and 26 surface mines producing 170 million short tons (see Table 13.2). These two states were chosen for comparison because

Table 13.2
State Coal Programs after SMCRA[a]

Region and State	Staff: Size	Staff: % Change	Federal % of Budget[f]	Mines: Number	Mines: % Change	Mines: Staff	New Permits /Staff[g]	Production: Total[b]	Production: % Change	Production: Mine
EASTERN STATES	1,528	176[d]		3,410	-42[d]	2.2	1.6	655,860	19[d]	192
Appalachian Coal Region	1,213	152	50	3,238	-43	2.7	1.8	499,628	12	154
Alabama	40	300	51	105	-56	2.6	5.2	27,992	24	267
Kentucky	466	133	50	1,099	-54	2.4	1.8	167,389	17	152
Maryland	11.5	28	50	31	-55	2.7	0.9	3,376	30	109
Ohio	116	45	51	184	-47	1.6	1.5	33,689	-28	183
Pennsylvania	367	267	50	681	-18	1.9	1.0	70,596	-16	104
Virginia	100	213	50	365	-51	3.7	2.0	43,006	21	118
West Virginia[c]	112	124	50	773	-28	6.9	3.3	153,580	41	199
Average of Appalachian States	173	158	50	463	-44	3.1	2.2	71,375	12	162
Interior Coal Region	316	338[d]	51	172	-9[d]	0.5	0.6	156,232	48[d]	908
Arkansas	9	800	50	7	-13	0.8	0.8	70	-86	10
Illinois	89	642	51	48	-17	0.5	0.4	59,267	0	1,235
Indiana	71	610	50	60	-3	0.8	1.3	33,641	34	561
Iowa	5	150	50	5	-50	1.0	0.9	430	-31	86
Kansas	8	60	50	5	25	0.6	0.5	856	79	171
Louisiana	6	--	50	2	--	0.3	0.1	2,983	--	1,492
Missouri	14.5	263	50	9	-31	0.6	1.0	3,378	-40	375
Oklahoma	47	370	52	21	-32	0.4	0.9	1,753	-39	83
Texas	66	136	50	15	275	0.2	0.1	53,854	389	3,590
Average of Interior States	35	379[d]	50	19	19[d]	0.6	0.7	17,359	38[d]	845
Average of all Eastern States	96	276[d]	50	213	-10[d]	1.7	1.3	40,991	26[d]	546

WESTERN STATES AND WESTERN COAL REGION	118.5	132[e]	77	105	8	0.9	0.3	301,375	281	2,870
Alaska	5	--	50	1	0	0.2	0.0	1,582	107	1,582
Colorado*	26	160	82	23	-30	0.9	0.3	17,123	108	744
Montana*	16	100	74	9	13	0.6	0.2	37,742	71	4,194
New Mexico*	11	57	64	8	60	0.7	0.2	23,702	170	2,963
North Dakota*	11	83	67	13	30	1.2	0.9	29,566	247	2,274
Utah*	20.5	583	84	21	5	1.0	0.2	20,102	189	957
Wyoming*	29	71	86	30	50	1.0	0.2	171,558	621	5,719
Average of Western States	17	176[e]	72	15	18	0.8	0.3	43,054	216	2,633

*Member of Western Interstate Energy Board.

[a] Data on coal production and mines is for 1989; staff size is based on 1990 responses to questionnaires supplemented by 1989 OSM staff figures (USDI 1990).

[b] Thousands of short tons.

[c] Data were obtained from telephone interview rather than questionnaire.

[d] Does not include Louisiana.

[e] Does not include Alaska.

[f] 50% of state regulatory program costs are paid through OSM for all lands; up to 100% of program costs are paid for administering an approved program on federal land.

[g] Number of new permits issued per staff member each year; includes mine and other facility permits; data from 1978-1986 and 1988, 1989 (USDI 1987, 1990).

Wyoming had a comprehensive environmental regulation program (including a competent regulatory staff) in place before SMCRA was enacted, and Utah did not. Some comparisons are also made between these western states and the eastern states of Ohio, Pennsylvania, Indiana, and Alabama.

WHAT PRECEDED SMCRA?

Most of the state agencies that regulate mining today evolved from smaller regulatory bodies that existed before SMCRA. A regional comparison reveals that eastern states have substantially more experience with coal regulatory programs than do western states. On average, eastern states have regulated coal mining and reclamation for about 35 years and western states for about 23 years.

The earliest mining regulatory programs were usually inspired by a single environmental issue (e.g., Indiana required tree plantings on coal spoil ridges in 1941) or were initially directed to health and safety issues (as in Oklahoma) (Hamm 1991; Norman 1991).

All of the eastern states, but only about one-third of the western states, had regulatory programs for ten or more years before their primacy programs were established. More than half of the western states began regulating coal mining between 1970 and 1975. However, years of regulatory experience do not provide any indication of the quality of the environmental protection afforded by the programs.

In comparing three eastern states, the Center for Science in the Public Interest (1975) concluded that "Kentucky's system of permit review is conducted in a deficient and hurried fashion by an untrained and understaffed section of DNREP (Department of Natural Resources and Environmental Protection). Inspectors—also involved in the permit review procedure—are overworked and incompetent."

While that same report criticized Pennsylvania for a general lack of inspections in the anthracite fields, it considered that state the most diligent (compared to Kentucky and West Virginia) in enforcement operations. Pennsylvania's law contained provisions for fines, bonding, and cease and desist orders that served to motivate coal operators to reclaim lands. The laws in Ohio and Pennsylvania had provisions for adequate bonding; Kentucky reported that the average amount of performance bond required would not cover reclamation costs (Thompson and Lindahl 1977). The differences in coal mining regulations among eastern states were so great that some elements of Pennsylvania's coal industry supported their state's participation in the drafting of SMCRA. They hoped that setting minimum national standards through federal regulations would require competitors in adjacent states to increase their environmental protection costs to levels comparable to those of Pennsylvania (Heine 1991).

In the West, Wyoming and Utah exemplify the range of programs administered under relatively new laws and regulatory agencies. Wyoming began to regulate mining under the Open Cut Land Reclamation Act of 1969. In 1973, the Wyoming legislature passed the Environmental Quality Act, a comprehensive environmental law that addressed all mining and other aspects of land quality, water quality, air quality, and solid waste management. The law was passed, among other reasons, in response to a real and anticipated upswing in mining (Diedrich 1973). At the same time, environmental groups were protesting in the Dakotas, Montana, and Wyoming because coal was being developed without comprehensive environmental impact assessment. By 1976, Wyoming had the staff and regulations to implement their new law. Wyoming's state program for implementation of SMCRA was approved in 1980, and a cooperative agreement with OSM to regulate mining on federal lands and federal minerals was signed in 1981.

Utah began to regulate mining in 1975, under the Utah Mined Land Reclamation Act. The law had an active sponsor in the legislature and support from an industry that could see the prospects of federal legislation in the near future. Administrative rules promulgated for noncoal mining remain essentially the same today except for removal in 1986 of a statutory exemption for mines under two acres and 500 tons per year. The Utah code was extensively revised in 1979 and 1981, and Utah's coal primacy program was approved in 1981. That state's cooperative agreement with OSM to regulate mining of federal lands and federal minerals was signed in 1987.

DEVELOPING STATE PRIMACY PROGRAMS

Two types of problems were encountered as states prepared to administer the new federal regulatory program. First, the states experienced problems in developing the infrastructure, rules, and procedures to administer a program. With any new program, these problems are likely to occur in states such as Utah and Alabama that had no experience with substantial environmental regulatory programs before SMCRA. Second, conflicts arose between the new federal agency and the state agencies. Conflicts may occur with all states, regardless of their regulatory history, because the federal agency itself lacks experience. They also occur when participants disagree on the role of the federal agency. These conflicts are more likely to arise in states where programs were well established before the federal initiative, since the state's role as the overseer of environmental protection has already been established and states may perceive federal activity in this area as preempting their programs.

Passage of SMCRA in August 1977 began the first phase of the state-federal regulatory apparatus, the Initial Program. Grants and state initial program rules were developed, to be followed by numerous consultations with the fledgling OSM. Initial program grants evolved into program development grants, and federal cost-sharing percentages declined from 80 percent of operating costs in the first year to 50 percent of costs by the third year. In the early stages of state primacy, many states found it difficult to envision any benefit from SMCRA. Despite federal monies, the additional costs to states were readily apparent. Staffing was an item of concern, even for states with well-developed programs since the need for specialized technical, financial, legal, enforcement, and administrative expertise was driven by the SMCRA ethic of "as stringent as" the federal regulations (section 505[b]). All states were faced with the need to pass new reclamation laws, amend old reclamation laws, or, at least, overhaul state regulations. While new state laws were, in many cases, more stringent than state legislatures had ever allowed, eventually even the mining industry supported the rule changes necessary for primacy. In the past, industry had to deal only with the state bureaucracy; they wanted that situation to return. In states with federal lands (e.g., Oklahoma and Utah), the industry has similarly supported development of cooperative agreements to reduce involvement with federal regulators and duplication of jurisdiction over mines with both federal and nonfederal land ownership.

While one of the goals of SMCRA was to standardize environmental controls among states, one of the short-term costs of primacy was a temporary loss of regulatory predictability. In the early years, industry was unable to anticipate what procedures would be used and could not be sure that the regulatory process would be fair and consistent. Many states changed their laws or regulations to deal with certain aspects of mining (e.g., highwalls and overburden) in a new way. Enforcement became more intense and involved different approaches. Industry appealed more decisions and requested hearings in the face of new requirements. Many of the state agencies were changing from small, informal groups, sometimes led by one powerful individual (as in Oklahoma), to large, complex groups with formal staff structures and procedures. Until the methods of operation under revised laws and rules were established, the process was unpredictable. Once state processes were established and the requirements of OSM oversight and budgeting were known, state programs once again became relatively predictable. In most cases, old personalized routines were traded for new standard guidelines, procedures, and rules. Regardless of stringency of the rules, predictability of the regulatory process benefits the coal industry by facilitating planning and acquisition of funds for mining projects.

While it provides little detail about reasons, this study, through its state survey, does document the early conflicts between states and OSM. Fifty percent of the states indicated that during the first half of their primacy period, their relationship with OSM was *acrimonious* or *barely tolerable*. The other half categorized the relationship as harmonious or *harmonious with rough spots*. More Appalachian states (83 percent) had a poor relationship with OSM in the early years, as compared to interior states (33 percent) or western states (43 percent).

CHANGES IN STAFF CAPABILITIES

More than a decade after SMCRA was enacted, measurable changes have occurred in state programs. Some of these can be attributed to the federal initiative.

Increasing Staff Size

Survey results indicate that before SMCRA, state regulatory agency staff sizes varied widely, ranging from no staff (no regulatory programs) to 200 employees. Agencies in the interior and western coal regions were all relatively small compared with those in the Appalachian region (Table 13.1). Presently, under SMCRA, state program staffs are larger, ranging from 5 to 500. In Appalachia average staff size increased 152 percent, and in western states it increased 132 percent. The interior region experienced the greatest change, increasing 338 percent.

To put these increases into perspective, both the number and average production of active mines presently and before SMCRA were compared. During the years since SMCRA was enacted, all three regions made gains in the personnel resources available to regulate mines.

In the Appalachian region, while staff sizes in state regulatory agencies have increased, the number of active mines has decreased by about 43 percent. Thus, the ratio of mines to state regulatory personnel has changed from about 12 to 1 before SMCRA to about 2.7 to 1 now. In the interior region, the number of mines decreased by 9 percent, bringing the pre-SMCRA ratio of 2.6 mines per state regulatory personnel to less than 1 to 1. Despite the decrease in the number of mines in the Appalachian and interior regions, their production has increased by 12 percent and 48 percent, respectively. In the western region, while the number of mines has increased overall by about 8 percent and production has tripled, staff increases have changed the ratio of mines to staff from 2 to 1 before SMCRA to less than 1 to 1 today (Table 13.2).

During the years since SMCRA was enacted, all three regions made gains in the personnel resources available to regulate mines. The difference between the Appalachian region and the interior and western regions may be attributed in part to the difference in size of mines between the two areas. The larger, more complex mines of the western and interior regions may require more staff per mine for technical reviews and inspections than are required for smaller Appalachian mines, regardless of the regulations in effect. Between 1978 and 1986, the average western region permit (including surface mines, underground mines, and other facilities such as preparation plants and roads) was for about eighteen hundred acres, the average Appalachian area permit was for 80 acres, and the average interior region coal permit was for about 400 acres (U.S. Department of the Interior 1987). Coal production per mine reflects the same trend (Table 13.2).

In the years since passage of SMCRA, federal money to support state employees has certainly been given to all primacy states. But did implementation of SMCRA represent a real increase in the capability of states to regulate coal mining? Was this staffing increase beyond that which could have been expected from the states themselves? And how has staff size increased relative to changes in mines and changes in regulatory requirements?

Federal Infusion of Resources or Expected State Growth?

Federal grants to state programs average 50 percent of state costs in both the Appalachian and interior regions. This is the basic amount authorized by SMCRA for program operations (section 705[a]). The western region receives about 72 percent of its funding for coal regulation from OSM (Table 13.2). Within the region, the annual grants for additional funding are awarded to states that have cooperative agreements for regulating mining on federal lands. These grants are to equal approximately the amount that the federal government would have expended for such regulation (section 705[c]). Staff increases do not, however, seem to be directly related to the level of funding. Neither the numerical increase nor the percentage increase in staff correlates with the percentage of budget contributed by OSM (Table 13.2).

While it is difficult to estimate what coal program staffing levels would be today without SMCRA, an analysis of the support that two states have given to regulation of noncoal mining can provide an indication.

The 1990 staffing level of 20.5 employees for the coal regulatory program in Utah represented an almost sixfold (583 percent) increase over a staffing level of three before SMCRA. Before SMCRA was enacted, Utah employed one noncoal staff member and four joint coal/noncoal staff. After passage of SMCRA, the state employed three noncoal staff, an increase of 300 percent, or about half that of the increase for coal personnel. In contrast, Wyoming's 1990 federally subsidized coal and state-financed noncoal staffing levels both increased about 70 percent over the pre-SMCRA staffing level, and the ratio of coal to noncoal workload and funding has consistently averaged about 60:40. While the legislatures of both states can be considered conservative in their financing of environmental control programs, the Wyoming legislature appropriates funding for about 16 noncoal staff, or five to six times the number of employees supported by Utah.

Since different commodities dominate the Utah and Wyoming minerals industries, and since these dominant commodities have changed within each state over the years, number of mines and quantity of production cannot be used to evaluate the potential increase in workload for minerals mining programs in these states. The total values of nonfuels mineral production for Utah and Wyoming indicate that activity in the mineral mining industry has generally followed the same pattern in these states (U.S. Bureau of Mines 1970–85). However, staffing of minerals regulatory programs has substantially increased only in Wyoming.

It can be argued, of course, that without 86 percent federal funding of the coal program, the Wyoming legislature would not have supported a 70 percent increase in its entire staff, and the total mine regulatory program would currently have a smaller overall staff, with more limited capabilities. But if the progress of noncoal mining regulation is any indication, it is likely that even without the federal financial support of SMCRA, the total mining regulatory program of Wyoming (and states like it) would have evolved into a program comparable to the existing one. It is also likely, based on its noncoal regulatory program, that without the federal stick and carrot, Utah's coal program would still be minimal.

What Has Increased Staffing Accomplished?

Regardless of differences among the regions in the types of mines, the resources needed to regulate mines today are not strictly comparable to those of pre-SMCRA days. Even if regulations had not changed, the general increase in mine size makes it likely that today's mines would require more state regulatory personnel per mine (for permitting, permit modifications, and inspection). The need for increased staffing is, however, probably due mostly to changes in regulatory requirements. In states like Wyoming or Pennsylvania, the basic regulatory requirements may not have changed significantly from pre-SMCRA to the present. But in states with minimal pre-SMCRA requirements (e.g., Utah and Alabama), today's mine permitting and enforcement activities require considerably more effort and resources.

Administrators from the Wyoming Land Quality Division note that even for a state with a good pre-SMCRA program, regulatory duties have increased significantly under SMCRA rules. They question the value of these additional duties. While they recognize that the number of mines has increased, they contend that additional staff is not required primarily because of that increase, nor because of a decrease in staff productivity. Rather, nearly all the increased staff that is supported by federal funding is needed to deal with additional paperwork and follow-up field work required by SMCRA regulations and OSM oversight policies. Regulatory staff increases in Wyoming have resulted in only a small decrease in the mine-to-staff ratio (1.2 to 1.0). Wyoming reports no *increase in effectiveness* of implementation of their program, only a moderate *improvement in on-the-ground compliance,* and no *increase in environmental protection* from the effects of coal mining as a result of SMCRA. In effect, additional federal money in Wyoming has, in part, replaced state money, and, in part, has paid for increased paperwork that has not resulted in significant improvement in environmental quality.

The situation in Pennsylvania is different. Despite the quality of Pennsylvania's pre-SMCRA program, implementation and funding of the federal law resulted in a significant improvement in the mine-to-staff ratio, from over 8 to less than 2 mines per staff member. On-the-ground program effectiveness improved from *moderately effective* to *highly effective.* In Pennsylvania, staff expansion was needed and was used effectively, primarily to increase the frequency of inspection, to allow review of more detailed permit applications, and to more closely consider the quality of overburden in order to predict potential post-mining pollution problems (Heine 1991).

Utah, which previously had minimal regulation, has received monies that would probably not have come from state sources. As a result of these funds, Utah reports *increased effectiveness* in implementing the state regulatory program, *improved on-the-ground compliance, increased envi-*

ronmental protection, and *greater public benefits from environmental protection.*

The evolution of the regulatory program in Alabama shows certain parallels with that of Utah. Before SMCRA was enacted, Alabama was the most poorly staffed coal regulatory program in the country, relative to the number of active mines (Table 13.1) and had a rudimentary regulatory program for all mining (Johnson 1991). Coal production has increased moderately, and the value of Alabama's nonfuels mineral production has followed the same general rises and declines as Utah and Wyoming (U.S. Bureau of Mines 1970–85).

With federal funding assistance, Alabama improved its pre-SMCRA mine-to-staff ratio of 24:1 to less than 3:1 in 1990. Starting out with inadequate staff and regulations, Alabama has reaped many of the same benefits from SMCRA that Utah did. Alabama's regulatory effectiveness has improved from *not effective* to *highly effective,* with *better on-the-ground compliance,* and *increased environmental protection.* In Alabama, as in Utah, little evolution has occurred in the noncoal program. The staff remains small (11 people, including those with responsibilities for health and safety issues), and the program operates with general rules under authority of the state's 1969 (noncoal) Surface Mining Act.

For states with minimal environmental regulations, a federally funded program can provide the stability in staff and funding necessary to assure environmental protection in the face of an indifferent or hostile state legislature. For states with well-developed laws and programs, the federal initiative may largely substitute federal dollars for state dollars. The impact on environmental protection in those states may be small, but it is important.

Staff Diversity and Quality

SMCRA rules and funding may have both necessitated and provided the funding for increased diversity of staff. The current minimum staff size of five, to regulate as few as one and as many as five mines, indicates that a certain minimum staffing level is desirable, if not necessary, to fulfill regulatory requirements of SMCRA.

States reported that current coal program staffs were usually more diverse in technical expertise than were pre-SMCRA staffs. Earth sciences, natural sciences, engineering, administration, and support people were represented on both pre- and post-SMCRA staffs. However, typical staffs of post-SMCRA agencies include an increased representation of geologists, hydrologists, biologists, and engineers. Representation in other areas of expertise such as administrators, support people, inspectors, soils specialists, agronomists, foresters, and lawyers is also reported to have increased in post-SMCRA state regulatory agencies. The survey data do not show the specific changes that occurred in each state agency, but staff diversification was noted as a particular benefit to Wyoming. While the essential functions and tasks of the coal program must be provided, federal funding may give state agencies the flexibility to hire specialists they would otherwise be hard pressed to justify to a state legislature. Wyoming's staff organization allows specialists hired for the coal program to function in the noncoal program as well.

While number and diversity of employees has increased with SMCRA, quality of staff has also improved. To maintain a high quality staff, state agencies must (1) attract educated and/or experienced people; (2) provide training to improve capabilities of staff; and (3) minimize loss of experienced employees.

Attracting experienced staff: To attract and keep experienced employees, agencies must provide adequate salary benefits, job security, and career satisfaction. State program salaries, benefits, and security depend largely on the general salary and benefits structure for state employees as a whole and on the whims of the state legislature, rather than on the federal funding process.

Until recently, new employees in Wyoming tended to be recent college graduates. However, as the energy industry has declined, industry and consulting firms have lagged behind government in terms of job security, and experienced employees have moved from industry into state government. In general, staff at all levels and fields of expertise in Utah are more capable (better educated and more experienced) than those employed prior to SMCRA or in the early years of SMCRA. As compared to the technical professional of the middle to late seventies, more employees in Wyoming and Utah have either one or two degrees in a technical specialty, plus one to five years of experience, either in another government agency, in industry, or as consultants.

Providing technical training: In states where entry level employees are not as well educated, OSM's technical training program has been important to staff development. In-service training, provided by OSM, comprises about a dozen courses, including technical training (hydrology, engineering principles, soils, and revegetation), inspection/enforcement training (data recording and hearing room skills), and administrative training (management and writing classes). In 1989, 1,250 participants attended 60 training sessions in 11 separate courses. State personnel accounted for 82 percent of participants (U.S. Department of the Interior 1990). Wyoming and Utah have criticized the technical training courses as being too elementary to improve staff skills, but they perceive the inspection/enforcement training as more valuable.

The Western Interstate Energy Board would like to see OSM resources redirected to provide more advanced training, to deal with important regional technical issues, and to support more state-directed research in technical problems (Larson 1991).

Minimizing staff losses: Turnover rates were generally judged to be *moderate* in all regions (15 percent reported *high* turnover rates, while 55 percent reported *moderate,* and 30 percent repeated *low* turnover rates). Moderate turnover is defined as occurring after a person has spent from two to five years in a particular job category. Many of the states lose staff to industry, consulting firms, and the federal government, all of which are thought to have higher salaries and better benefits for comparable work. Job advancement (including pay increases) and the opportunity for different, more varied experience were cited by administrators as the main reasons for employee turnover. Other reasons, cited by past employees, include dissatisfaction with the repetitiveness of the regulatory process and with the work atmosphere.

While repetitiveness cannot be attributed solely to the federal program, employees (past and present) in both Wyoming and Utah commented that OSM's permitting style compounds the repetitive aspects of the work. The work atmosphere can be affected by the nature and the quality of the relationship between state and federal regulators. In many states, job satisfaction of administrators improved with SMCRA, primarily because political influence over their decisions declined. This removal of political influence could have a positive effect on the entire regulatory organization (Heine 1991). Dissatisfaction with OSM oversight, described by some long-time state employees as "oppressive," counteracts this effect in some states.

Utah noted that employee turnover was a problem for industry during the state's intense SMCRA repermitting effort in the early days of state primacy. Coal companies would reach an accord with a staff member on compliance or design. That staff member might then accept other employment and negotiations would have to begin again. Guidelines have since been adopted, but a team approach to technical problem solving has also helped maintain continuity in dealing with compliance and mine plan review issues. Utah, as well as other states, is currently dealing with the continuity problem by rewriting permanent program rules. The objective is to provide logical and streamlined rules that are easier to follow.

FEDERAL-STATE COOPERATION OR CONFLICT IN ENVIRONMENTAL PROTECTION

Benefits of the Federal Program

In general, state administrators indicated that the primacy program has resulted in increased environmental protection from the impacts of coal mining. The administrators were asked to rate the benefits gained through their state's primacy program. Ratings were made on a scale of 1 to 7, with 1 being high, and states were given an opportunity to choose either *no benefits* or *other* as categories for rating. Twelve of the states ranked the category *increased environmental protection from the impacts of coal mining* as either 1 or 2. Thirteen states, including six states with a high percentage of federal funding, rated *more funding to carry out the regulatory program* as 1 or 2. It is not completely clear whether increased environmental protection should be attributed to better laws (i.e., more stringent or more specific), or if it is also a function of more funding for states to implement environmental control programs. Several states that receive only 50 percent administrative grants did, however, rank *more funding* as low and *increased environmental protection* as high on their list of benefits. As discussed above, most of the states also indicated that their ability to implement regulations improved. *Better on-the-ground compliance* was rated as only moderately important by twelve states and as high by only five.

Additional responses, in order of importance, were as follows: (4) *improved post-mining land use potential*; (5) *more public benefits from increased environmental protection*; (6) *greater state/federal cooperation;* and (7) *no benefits or other.* There appeared to be little regional difference in how states observed and ranked the benefits of the state primacy program.

In conservative, industry-oriented states, one of the benefits of federal laws and oversight is the ability state regulators have to cite federal minimum regulatory standards when industry or the legislature attempt to dilute the effectiveness of environmental protection programs. While Wyoming has been forced to use this tactic on occasion, the strength of its pre-SMCRA Environmental Quality Act and subsequent additions to law and regulation for in-situ mining not covered by SMCRA suggest that the state does not need the threat of federal intervention to adequately protect its environment. Utah, on the other hand, has never passed environmental protection legislation or developed regulations that are more stringent than the federal laws. On the contrary, a Utah law prohibits state agencies from adopting regulations more stringent than applicable federal regulations.

The legislature did, however, specify in the Utah Code that the Division of Oil, Gas and Mining (UDOGM) could adopt more stringent regulations when they could demonstrate that the federal regulations were inadequate. This language applies only to coal regulations.

Staff in Utah's coal program consistently (and informally) cite the federal regulations and federal oversight to persuade coal operators that political appeals at the state level would be futile. The conservative political atmosphere of the state can be illustrated by UDOGM's attempt in 1987 to regulate a previously unregulated area of noncoal mining (limestone mining at a sand and gravel operation). The legislature countered the agency's action by excluding limestone mining from regulatory control.

SMCRA has also helped Alabama maintain regulatory control of coal mining. Alabama has a sunset law that requires a review every four years to determine the need for state agencies. The Alabama Surface Mining Commission would have been eliminated in the mid-1980s had SMCRA not mandated a specific level of regulatory control and provided 50 percent of the commission's funding. Since the requirements of SMCRA were explained to the state legislature, they have taken a hands-off attitude toward the commission (Johnson 1991).

Conflict over State-Specific Programs

Interviews with state program administrators provide some insight into the problems associated with the federal program. Wyoming had a comprehensive environmental regulation program before the federal program was imposed. Thus from Wyoming's perspective, SMCRA, its specific regulatory direction, and the OSM oversight that has resulted, constitute unnecessary interference by bureaucrats who do not understand the specific problems of the West and who are unwilling or unable to make decisions. Wyoming believes it is in the spirit of SMCRA and in accord with the policy of the Bush administration for states to have the freedom to apply the basic principles of SMCRA, as appropriate to the state. They cite SMCRA section 101(f) and Executive Order 12612 on federalism. The latter requires federal agencies to "refrain, to the maximum extent possible, from establishing uniform standards for programs and, when possible, defer to the states to establish standards." With respect to national policies administered by the states, the "national government should grant the states the maximum administrative discretion possible" (52 FR 41685).

Wyoming's strong statement was echoed by administrators from many states, both East and West. A majority of those surveyed indicated that they had tried to modify their coal programs to fit particular state environmental concerns or physical conditions that affect reclamation techniques or success. Questionnaire choices ranged from OSM being *very receptive and helpful* to OSM *thwarted most efforts*. Sixty-three percent of the respondents believed that OSM had *thwarted most efforts,* 16 percent said OSM was *receptive,* and 21 percent indicated OSM efforts fell between these two extremes. A recurring theme in questionnaire responses and follow-up interviews was that it is more difficult now to negotiate state-specific regulations with SMCRA than it was during initial development of state programs. One state administrator conjectured that because the 1983 OSM rules revisions provided states with more room for interpretation, little consideration would be given to additional requests for state-specific changes, despite the appropriateness of some requests (Johnson 1991). One former state and federal administrator speculated that some states are currently seeking exceptions for practices and procedures specifically contrary to SMCRA provisions (Heine 1991).

The survey showed no differences among respondents, on a regional basis, in regard to this question. The Western Interstate Energy Board (WIEB) has, however, taken particular issue to what it terms OSM's "drive toward national uniformity in implementation of the Act." This, the board says, "is leading to ever more strange and counter-environmental actions in the West" (Larson 1991).

SMCRA and its implementing regulations are generally seen as eastern in orientation. Exceptions are some basic provisions, such as reclamation bonding, and a few special provisions that specify lands west of the 100th meridian (e.g., alluvial valley floors, section 510[b][5]), or spell out revegetation standards for areas of minimal rainfall (section 515[b][20]). While specific federal regulations may address eastern problems, at the same time they force western states to apply unnecessary, and sometimes counterproductive, measures.

Wyoming believes that OSM has *thwarted most efforts* to modify the coal program to fit particular state environmental concerns, as required by SMCRA (section 101[f]), but Utah administrators say that OSM has been *skeptical but cooperative* on requested rule changes. OSM has not, however, allowed Utah any discretion on enforcement issues.

In some cases, OSM is willing to allow state-specific practices if those practices are specifically authorized through program amendments. However, it will not allow the states the flexibility to permit alternative practices on a case-by-case basis.

STATE-FEDERAL WORKING RELATIONSHIPS

Appropriate regulations and an adequate number of competent staff are essential elements for a successful environmental control program. For a state program, these elements may be sufficient. However, for a federal program that involves state primacy, the relationship between state and federal officials can be crucial to the program's success. This relationship can affect staff morale and influence turnover, and it can affect the ability of states to modify regulations. It can also facilitate or obstruct the day-to-day implementation of regulations, thus affecting not only protection of the environment, but also the industry regulated.

Federal-State Relationships

As indicated earlier, half the states had a poor relationship with OSM in the early days of primacy (*acrimonious* or *barely tolerable*); the other half had a relationship that was generally good (*harmonious* or *harmonious with rough spots*). The relationship improved for 57 percent of the states and remained the same for another 33 percent. Currently, about 85 percent of the states indicate that their relationship with OSM is generally good (63 percent reported *harmonious with rough spots* and 22 percent reported *harmonious*).

Wyoming provides insight into the source of current rough spots by comparing the regulatory style of OSM to that of the state's Land Quality Division. OSM regulates primarily "by design," while the Land Quality Division regulates "by performance resulting from design."

While OSM changed its official policy in the early 1980s to favor the performance approach to permitting, its interactions with the states do not necessarily support this policy. Recent proposals for best technological controls for sediment, for example, revert to the design approach, despite the improvements in state staff capabilities to deal with technical mining problems.

Wyoming administrators feel that regulation by performance is more effective and that their staff organization complements this style of regulation. Inspection/enforcement staff are distributed throughout the coal region with the authority to make on-the-ground changes. On the other hand, Utah's organization, with staff and most inspectors centrally located—hours from the mine sites—may not be as conducive to the performance type of regulatory style. Certainly, from the federal perspective, oversight activities are more easily accomplished when all states operate under the same procedures and regulations. With the OSM style of regulation, the need for a strong institutional memory (which is facilitated by a stable staff) is not as critical since all aspects of the mine operation are recorded in detail in the permit document. Wyoming's answer to creating a strong institutional memory in the face of high staff turnover was to develop and use state guidelines that do not have to be approved by OSM.

Wyoming is not alone in criticizing OSM for being more concerned about how things appear on paper than how they work in the coal field. As one Montana commissioner described the situation, "OSM, primarily in response to Congressional pressure, is far more concerned with whether or not the paper is moving correctly than with whether or not reclamation is working" (Hemmer 1985). More recently, WIEB charged that there is an attitude of confrontation between OSM and the western states "when western States see OSM focusing on process questions rather than fundamental environmental protection questions. . . . The interaction between OSM and western States needs to be redirected away from questions of process and toward those areas where action is needed to protect the environment" (Larson 1991).

State-to-State Communication

While relations between the states and OSM have been strained, communication among the states has increased and become more open. Professional groups play a key role in enhancing communication in the technical, administrative, and political realms. As technical professionals and lawyers became more involved in the regulatory process, their agendas slowly began to include topics specifically related to reclamation and SMCRA. Professional associations began to feature their own technical recognition of mining and reclamation, and that led to more sharing of information among individuals who represented the various states. Likewise, organizations such as reclamation associations, the Western Interstate Energy Board, Interstate Mining Compact Commission, task groups commissioned by OSM itself, Bureau of Land Management, and academia include many disciplines and draw together the people who do the day-to-day work of state regulatory programs. Over the years, both the frequency and intensity of these informal contacts have increased, and it appears they will continue to thrive, forming an unofficial network among people, the states, and OSM.

CONCLUSION

The state agencies responsible for controlling the impacts of coal mining have certainly evolved since SMCRA was implemented. Staff size has increased universally as federal funding has replaced and supplemented state monies. Technical staffs have diversified in most states. Improvement in staff competence through in-service training has depended largely on the education/experience level of an agency's staff. Information exchange and cooperation among states has also developed. With these changes, many states have come to believe that regulations are being more effectively applied, resulting in better environmental protection.

But SMCRA has brought problems as well as benefits. Some internal state agency problems have not been diminished with implementation of a federal program and were exacerbated through poor federal-state relations. Various aspects of OSM oversight are considered to be annoyances, as well as leading to misallocation of resources; others are considered to be counterproductive. The continuing drive to national uniformity in implementing SMCRA is seen as a serious problem, particularly in the West. OSM's attempt to impose uniformity seems to have affected states, regardless of the level of sophistication of their pre-SMCRA programs. Uniformity may be the problem with the most significant environmental implications. Other problems, such as lack of regulatory predictability, may be unavoidable as new regulatory requirements are implemented, regardless of whether these requirements are federally or state instigated.

It is undeniable that the overall ability of states to protect the environment is better today than it would have been without the federal program. The progress of states with minimal pre-SMCRA programs demonstrates the importance of a federal program to set a minimum standard and provide at least the minimum resources to implement that standard. States with significant pre-SMCRA regulatory programs show evidence that their commitment to environmental protection would have continued without either the carrot of federal monies or the stick of federal regulations. But despite federal-state tensions and specific program problems, the capability of those states to control impacts of mining has certainly not declined with SMCRA.

The decade of the eighties has seen state capabilities, including both staff and regulations, develop to a common baseline for environmental protection. Further improvements will require federal-state cooperation to make all aspects of the national program responsive to state-specific needs and to make full use of state regulatory capabilities.

REFERENCES

Backs, Jean. 1991. Ohio Division of Reclamation. Personal communication with Kathryn Mutz.

Center for Science in the Public Interest. 1975. *Enforcement of Strip Mining Laws*. Report of a survey by the Center for Science in the Public Interest.

Diedrich, Roger A. 1973. "The Mineral Industry in Wyoming." *Minerals Yearbook.*

Energy Information Administration. 1989. *Coal Production 1989*. Washington, D.C.: U.S. Department of Energy, National Energy Information Center. DOE/EIA-0118(89).

"Excavating the History of Mined Land Reclamation in Indiana." 1990. *Dragline* (Indiana Division of Reclamation) January: 4–5.

Hamm, James. 1991. Deputy Director for Coal, Oklahoma Department of Mines. Personal communication with Kathryn Mutz.

Heine, Walter N. 1991. Walter N. Heine Associates, Inc. and former Director of OSM. Personal communication with Kathryn Mutz.

Hemmer, Dennis. 1985. "Regulatory Framework Governing Coal Exploration and Mining." *Montana Coal Forum*. Montana Bureau of Mines and Geology Special Publication 93.

Johnson, Randall. 1991. Director of Alabama Surface Mining Commission. Personal communication with Kathryn Mutz.

Larson, Douglas. 1991. Letter to Jennifer Salisbury, Deputy Assistant Secretary, Lands and Mineral Management, Department of the Interior, January 18.

Norman, Julie. 1991. Public Affairs, Indiana Division of Reclamation. Personal communication with Kathryn Mutz.

Summers, Rick. 1990. Hydrologist, Utah Division of Oil, Gas and Mining. Personal communication with Kathryn Mutz.

Thompson, Duane, and David Lindahl. 1977. *State Surface Mining Laws: A Survey, a Comparison with the Proposed Federal Legislation, Background Information*. Prepared for the Senate Committee on Energy and Natural Resources. Washington D.C.: Environmental and Natural Resources Policy Division, Congressional Research Service, Library of Congress.

U.S. Bureau of Mines. 1970–85. *Minerals Yearbook*. Washington, D.C.: U.S. Government Printing Office.

U.S. Department of the Interior. 1990. *Office of Surface Mining Annual Report Fiscal Years 1988 and 1989*. Washington, D.C.: U.S. Government Printing Office.

———. 1987. *Surface Coal Mining Reclamation: 10 Years of Progress, 1977–1987*. Washington, D.C.: U.S. Government Printing Office.

Table of Cases

Select Bibliography

Abrams, Dennis. 1982. "The Federal Surface Mining Control and Reclamation Act of 1977—First to Survive a Direct Tenth Amendment Attack." *West Virginia Law Review* 84: 1069.

Advisory Commission on Intergovernmental Relations. 1984. *Regulatory Federalism: Policy, Process, Impact, and Reform*. Washington, D.C.: U.S. Government Printing Office.

Barry, Hamlet J. 1981. "The Surface Mining Control and Reclamation Act of 1977 and the Office of Surface Mining: Movable Targets or Immovable Objects?" *Rocky Mountain Mineral Law Institute* 4: 117.

Beier, Ann E., and James M. McElfish. 1989. "The Need for Federal Oversight of Alternative Bond Systems Under SMCRA." *Policy Studies Review* 9: 132–42.

Berman, Paul. 1980. "Thinking About Programmed and Adaptive Implementation: Matching Strategies to Situations." In H. M. Ingram and D. E. Mann (eds.), *Why Policies Succeed or Fail*. Beverly Hills, Calif.: Sage Publications.

Casenote. 1990. "Meeting the Requirements of Standing: A Framework for Environmental Interest Groups: Lujan v. National Wildlife Federation, 110 S. Ct. 3177." *Hamline Law Review* 14: 277.

Conrad, Gregory E. 1987. "The Applicant Violator System Under the Surface Mining Control and Reclamation Act of 1977: Constitutional Concerns." *Eastern Mineral Law Institute* 8: 10.1–10.49.

Dernbach, John C. 1986. "Pennsylvania's Implementation of the Surface Mining Control and Reclamation Act: An Assessment of How 'Cooperative Federalism' Can Make State Regulatory Programs More Effective." *University of Michigan Journal of Law Reform* 19: 903.

Dragoo, Denise A. 1990. "Subsidence Controls and Section 522 of the Surface Mining Control and Reclamation Act." University of Utah, *Journal of Energy, Natural Resources and Environmental Law* 2(1): 31–48.

Galloway, L. T., and T. J. FitzGerald. 1985. "Abuse of the Surface Mining Act: A Continuing Story." *West Virginia Law Review* 87(3): 627–47.

Gellhorn, Ernest, and Ronald Levin. 1990. *Administrative Law and Process*. 3d ed. St. Paul, Minn.: West Publishing Co.

Harris, R. 1989. "Federal-State Relations in the Implementation of Surface Mining Policy." *Policy Studies Review* 9: 69–78.

Harris, Richard A. 1985. *Coal Firms and the New Social Regulation*. Durham, N.C.: Duke University Press.

Hawkins, Robert B., Jr. 1990. "Pre-emption: The Dramatic Rise of Federal Supremacy." *Journal of State Government* 63: 10–14.

Hays, S. 1987. *Beauty, Health, and Permanence*. New York: Cambridge University Press.

Irwin, Will A. 1988. "Federal Administrative Review Under the Surface Mining Control and Reclamation Act: Annotated Procedural Guide." *Journal of Mineral Law & Policy* 3: 417.

Kirst, Michael, and Richard Jung. 1982. "The Utility of a Longitudinal Approach in Assessing Implementation: A Thirteen-Year View of Title I, ESEA." In Walter Williams and others, *Studying Implementation: Methodological and Administrative Issues*. Chatham, N.J.: Chatham Publishers, Inc.

Lindblom, Charles. 1968. *The Policy Making Process*. Englewood Cliffs, N.J.: Prentice-Hall.

Lowi, Theodore J. 1979. *The End of Liberalism*. 2d ed. New York: W. W. Norton and Co.

McElfish, James, and Ann Beier. 1990. *Environmental Regulation of Coal Mining*. Washington, D.C.: Environmental Law Institute.

Mazmanian, Daniel A., and Paul A. Sabatier. 1983. *Implementation and Public Policy*. Glenview, Ill.: Scott, Foresman and Co.

Nakamura, Robert T., and Frank Smallwood. 1980. *The Politics of Policy Implementation*. New York: St. Martin's Press.

Nathan, Richard P., and Fred C. Doolittle. 1987. *Reagan and the States*. Princeton, N.J.: Princeton University Press.

Quinn, Harold. 1989. "Coal Resource Development and Land Use Planning: The Demands of SMCRA." *Natural Resources and Environment* 3(4): 24.

Rasnic, Carol C. 1982. "The Conflict Between the Environmentalists and the Surface Mining Industry: Federal Legislation and Litigation." *Northern Kentucky Law Review* 9: 427.

Rochow, James. 1978. "The Far Side of Paradox: State Regulation of the Environmental Effects of Coal Mining." *West Virginia Law Review* 81: 559.

Rosenbaum, Walter A. 1987. *Energy Politics and Public Policy*. Washington, D.C.: Congressional Quarterly.

Squillace, Mark. 1985. "Cooperative Federalism Under the Surface Mining Control and Reclamation Act: Is This Any Way to Run a Government?" *West Virginia Law Review* 87: 687.

Sunstein, Cass R. 1983. "Deregulation and the Hard-Look Doctrine." *The Supreme Court Review Annual*. 177–213.

U.S. General Accounting Office. 1989. *Surface Mining: Operation of the Applicant Violator System Can Be Improved, GAO/AFMD-89–31*. Washington, D.C.: U.S. Government Printing Office.

———. 1987. *Surface Mining: State and Federal Use of Alternative Enforcement Techniques, GAO/RCED-87–160*. Washington, D.C.: U.S. Government Printing Office.

Van Meter, Donald S., and Carl Van Horn. 1975. "The Policy Implementation Process: A Conceptual Framework." *Administration and Society* 6: 445–88.

Vestal, Theodore M. 1988. "Federal Administration of the Surface Mining Control and Reclamation Act of 1977 in Oklahoma." *Publius* 18 (Winter): 45–60.

———. 1988. "First Decade of the Implementation of the Surface Mining Control and Reclamation Act in Oklahoma." *Tulsa Law Journal* 23: 593.

Vietor, R. 1980. *Environmental Politics and the Coal Coalition*. Dallas: Texas A & M University Press.

Wassenberg, Pinky S. 1986. "Implementation of Intergovernmental Regulatory Programs: A Cost-Benefit Perspective." In J. Edwin Benton and David R. Morgan (eds.), *Intergovernmental Relations and Public Policy*. Westport, Conn.: Greenwood Press.

Welborn, David M. 1988. "Conjoint Federalism and Environmental Regulation in the United States." *Publius* 18: 27–43.

Williams, Walter. 1982. "The Study of Implementation: An Overview." In Walter Williams and others, *Studying Implementation: Methodological and Administrative Issues*. Chatham, N.J: Chatham Publishers, Inc.

Index

List of Contributors

ROBERT E. BECK, School of Law, Southern Illinois University at Carbondale.

GREGORY E. CONRAD, executive director, Interstate Mining Compact, Hendron, Virginia.

RONALD W. DANIELS, Division of Oil, Gas and Mining, Utah Department of Natural Resources, Salt Lake City.

JOHN C. DERNBACH, special assistant, Pennsylvania Department of Environmental Resources, Harrisburg.

UDAY DESAI, professor, Department of Political Science, Southern Illinois University at Carbondale.

DENISE A. DRAGOO, attorney, Fabian, Clendenin, Attorneys at Law, Salt Lake City, Utah.

DAVID HEDGE, associate professor, Department of Political Science, University of Florida, Gainesville.

JAMES M. MCELFISH, JR., Environmental Law Institute, Washington, D.C.

DONALD C. MENZEL, professor, Department of Government and International Affairs, University of South Florida, Tampa.

PATRICIA METZ, analyst, General Accounting Office, Washington, D.C.; Ph.D. candidate in Political Science, West Virginia University, Morgantown.

RICHARD O. MILLER, chief, planning and analysis staff, Office of Surface Mining, U.S. Department of the Interior.

LISA J. MORRIS, general counsel, Division of Reclamation, Ohio Department of Natural Resources, Columbus, Ohio.

KATHRYN M. MUTZ, School of Law, University of Colorado, Boulder.

HAROLD P. QUINN, JR., vice president and counsel, National Coal Association, Washington, D.C.

DENISE SCHEBERLE, assistant professor, Department of Political Science, University of Wisconsin at Green Bay.

MICHAEL J. SCICCHITANO, associate professor, Department of Political Science, University of Florida, Gainesville.

About the Editor

UDAY DESAI is Director of the Master of Public Administration Program and Professor of Political Science at Southern Illinois University, Carbondale. He specializes in environmental policy, public policy implementation, and natural resources management, and has written extensively on environmental policy implementation.

Policy Studies Organization publications issued with Greenwood Press/Quorum Books

Public Administration and Decision-Aiding Software: Improving Procedure and Substance
Stuart S. Nagel, editor

Outdoor Recreation Policy: Pleasure and Preservation
John D. Hutcheson, Jr., Francis P. Noe, and Robert E. Snow, editors

Conflict Resolution and Public Policy
Miriam K. Mills, editor

Teaching Public Policy: Theory, Research, and Practice
Peter J. Bergerson, editor

The Reconstruction of Family Policy
Elaine A. Anderson and Richard C. Hula, editors

Gubernatorial Leadership and State Policy
Eric B. Herzik and Brent W. Brown, editors

Public Policy Issues in Wildlife Management
William R. Mangun, editor

Health Insurance and Public Policy: Risk, Allocation, and Equity
Miriam K. Mills and Robert H. Blank, editors

Public Authorities and Public Policy: The Business of Government
Jerry Mitchell, editor

Technology and U.S. Competitiveness: An Institutional Focus
W. Henry Lambright and Dianne Rahm, editors

Using Theory to Improve Program and Policy Evaluations
Huey-tysh Chen and Peter H. Rossi, editors

Comparative Judicial Review and Public Policy
Donald W. Jackson and C. Neal Tate, editors

www.ingramcontent.com/pod-product-compliance
Ingram Content Group UK Ltd.
Pitfield, Milton Keynes, MK11 3LW, UK
UKHW041639031026
15587UKWH00001B/391

9 780313 286988